Build Your Own Plug-In Hybrid Electric Vehicle

TAB Green Guru Guides

Consulting Editor: Seth Leitman

Renewable Energies for Your Home: Real-World Solutions for Green Conversions by Russel Gehrke

Build Your Own Plug-In Hybrid Electric Vehicle by Seth Leitman

Build Your Own Electric Motorcycle by Carl Vogel

Build Your Own Plug-In Hybrid Electric Vehicle

Seth Leitman

New York Chicago San Francisco
Lisbon London Madrid Mexico City
Milan New Delhi San Juan
Seoul Singapore Sydney Toronto

The McGraw-Hill Companies

Library of Congress Cataloging-in-Publication Data

Leitman, Seth.
Build your own plug-in hybrid electric vehicle / Seth Leitman.
p. cm.
Includes index.
ISBN 978-0-07-161473-3 (alk. paper)
1. Hybrid electric vehicles. 2. Automobiles, Home-built. I. Title.
TL221.15.L45 2009
629.22'93—dc22 2009016405

McGraw-Hill books are available at special quantity discounts to use as premiums and sales promotions, or for use in corporate training programs. To contact a representative, please e-mail us at bulksales@mcgraw-hill.com.

Build Your Own Plug-In Hybrid Electric Vehicle

1 2 3 4 5 6 7 8 9 0 DOC/DOC 0 1 5 4 3 2 1 0 9

ISBN 978-0-07-161473-3
MHID 0-07-161473-7

The pages within this book were printed on acid-free paper containing 100% postconsumer fiber.

Sponsoring Editor
Judy Bass

Editorial Supervisor
Stephen M. Smith

Production Supervisor
Pamela A. Pelton

Project Manager
Patricia Wallenburg, TypeWriting

Copy Editor
Alice Manning

Proofreader
Paul Tyler

Indexer
Karin Arrigoni

Art Director, Cover
Jeff Weeks

Composition
TypeWriting

About the Author

Seth Leitman is the president and managing member of the ETS Energy Store, LLC (www.etsenergy.com), which offers consulting services to companies and individual families on organic, natural, and sustainable products. He is also a writer for the Planet Green Web site, and the Consulting Editor for the McGraw-Hill Green Guru Guides. Mr. Leitman runs the blog www.greenlivingguy.com, and is the co-author of the bestselling *Build Your Own Electric Vehicle*, Second Edition, published by McGraw-Hill. For more information, visit www.sethleitman.com.

Contents

Preface

I am an electric vehicle purist and the electric vehicle's time has come. However, I do realize that we need to get from Point A to Point B, and a plug-in hybrid electric vehicle is the logical next step for the millions of people who own hybrid cars or who want an extremely efficient car and realize that it is quicker to build than wait for the car companies.

The plug-in hybrid electric vehicle (PHEV) movement has broadened to multiple levels of the public debate. Al Gore and Leonardo DiCaprio have recently made movies about the need to assist the environment and how oil and energy have created the global warming problems that our world currently faces. Al Gore won a Nobel Peace Prize in 2007 for his discussion and call to action about global warming/climate change. Oil has flirted with prices of over $140 per barrel and $4.50 per gallon.

Well, PHEVs solve a lot of problems quickly. PHEVs bypass high energy prices and they cost pennies to charge. PHEVs have zero tailpipe emissions when running all electric.

While they charge up on electricity from power plants, they can also be charged by electricity from solar, wind, and any other renewable resource. Also, if you compare emissions from power plants for every car on the road with gasoline emissions, the electric component of the cars is always, *always* cleaner. In addition, as power plants get cleaner and reduce emissions, PHEVs will only get cleaner.

Plug-in hybrid electric cars also help develop the economy. We all know that we need to increase the expansion of electric cars. Hybrid electrics, plug-in hybrids, and low-speed vehicles all increase electric transportation. We as a country—no, we as a world—are increasing our involvement in this industry. From China and India, to Great Britain and France, and back here in the United States, electric transportation can create a new industry that will increase our manufacturing sector's ability to build clean, efficient cars. I recently spoke with an owner of an electric car company who said that the UAW was more than excited about building electric cars since the traditional car companies were leaving Detroit in single file. So a switch to electric cars can only increase domestic jobs in the United States and help our economy. Our world is dependent on fossil fuels from countries whose actions have not predominantly been in the best financial interests of the United States. Sending over billions of dollars to countries that are politically unstable and/or antagonistic to Western nations is a national security issue for all of us. Another way to ask the question is: Should we be sending more money to Iran and Venezuela, or should we keep it in our own pockets? What is the point? There is no point. That is why I believe in a pollution-free, oil-free form of transportation.

Who I Am

I am the managing member of a store called the ETS Energy Store, LLC (www .etsenergy.com), which sells organic, natural, and sustainable products for businesses, homes, and families. In addition, I run a blog called www .greenlivingguy.com that discusses sustainable living, energy efficiency and electric cars—green living! I also write for the Planet Green Web site (http://planetgreen. discovery.com) on clean transport, as well as for Greenopia (www .greenopia.com) on everything from electric cars to insulation. Thanks, Starre Vartan, for that last one!

In addition, I recently wrote the book *Build Your Own Electric Vehicle*, Second Edition, with Bob Brant. The success of this book has led me to become a consulting editor to McGraw-Hill on a line of books called the *Green Guru Guides*. This book is part of that series!

I am a New Yorker who rode the electric-powered subway trains. In fact, when I worked for the New York Power Authority, which powers those subways, I gained a new appreciation for electric transportation every time I took the train into and around New York City. My interest in renewable energy and energy efficiency, however, began in graduate school at the Rockefeller College of Public Affairs and Policy in Albany, New York, where I received a Master of Public Adminstration. I concentrated on comparative international development, which focused on the World Bank and the International Monetary Fund.

After I read about the World Bank's funding of inefficient and environmentally destructive energy projects, such as coal-burning power plants in China and dams in Brazil that had the potential to destroy the Amazon, I decided to take my understanding to another level. For my master's thesis I interviewed members of the World Bank, the International Monetary Fund, and the Bretton Woods Institution. I was fortunate enough to be able to ask direct questions of the project managers who oversaw billions of dollars of funding that went to China to build coal-burning power plants. I asked them how the Bank could fund an environmentally destructive energy project with no traps or technologies to recapture the emissions and use that energy. The answers were not good. But since I researched the Bank, attention to environmental issues has expanded by leaps and bounds, and the Bank is starting to work toward economic and environmental efficiencies. While it still has a lot of work to do, it's clear that progress is possible.

This passion to understand how organizations could create environmentally positive, energy-efficient economic development programs (all in one) led me to work for the New York State Energy Research and Development Authority (NYSERDA). While at NYSERDA, I was the lead project manager for the U.S. DOE Clean Cities Program (a grass roots–based program that develops alternative-fueled vehicle projects across the country) for five of the seven Clean Cities in New York. I was also the lead project manager for the Clean Fueled Bus Program for the Clean Air/Clear Water Bond Act, which provided over $100 million in incremental cost funding for transit operators to purchase alternative-fueled buses. When I funded programs and realized the benefits of electric cars or hybrid electric buses relative to their counterparts, I was transformed. I saw that electric transportation was the way to go.

I was so in love with electric transportation that I went to work for the New York Power Authority's electric transportation group. In total, I helped to bring more than 3,500 alternative-fueled vehicles and buses into New York. I was the market development and policy specialist for the New York Power Authority, the nation's largest publicly owned utility. I worked on the development, marketing, and management of electric and hybrid vehicle programs serving the New York metropolitan area. I developed programs that expanded the NYPA fleet from 150 to over 700 vehicles, while enhancing public awareness of all the programs.

I was the lead manager of the NYPA/TH!NK Clean Commute Program™, which under my leadership expanded from 3 to 100 vehicles. This program was the largest public/private partnership of its time. I secured and managed a $6.5 million budget funded by the federal government, project partners, participants (electric vehicle drivers), and Ford Motor Company. I developed an incentive program that offered commuters who drove electric cars up-front parking at train stations with electric charging stations. We provided insurance rebates and reduced train fares. This program secured media coverage in *USA Today*, the Associated Press, Reuters, *The New York Times*, CNN, *Good Morning America*, the *Today* show, and other media sources. To date, this was the largest electric vehicle station car program in the world.

As seen in Figures 1, 2, and 3, I was able to have 20 cars leased in Chappaqua, New York. It was one of the most successful train stations (excluding Huntington and Hicksville in Long Island, New York) for the program. Figure 1 is a great overhead shot of the cars lined up in the areas we set up right next to the station to give prime parking as an incentive for participating. Figure 2 shows one of the

Figure 1 An electric car dream. TH!NK City electric cars at Chappaqua train station in the NYPA/TH!NK Clean Commute Program. Photo courtesy of Town of New Castle.

Figure 2 TH!NK City cars charging with AVCON chargers at Chappaqua train station. Photo courtesy of Town of New Castle.

charging stations up close; it used an AVCON charger with an overhead light. The station connector cables were designed to be like a regular gas station (thanks to Bart Chezar, former manager of the Electric Transportation Group; Sam Marcovicci, who was the electric charger specialist for NYPA at the time; and ETEC out in Arizona, listed in Chapter 12). Figure 3 shows how close the charging stations were to the station, since you can see the platform and station name on the platform behind the row of TH!NK City EVs. It was a great program.

Figure 3 Look how close the electric cars were to the station platform. Talk about incentives! Photo courtesy of Town of New Castle.

On a related note, I am so glad to see that TH!NK is reemerging in the international automotive marketplace and look forward to its return to the United States.

I led and worked with multiple state and local agencies to place more than 1,000 GEM and TH!NK Neighbor low-speed vehicles for their respective donation programs to meet zero-emission vehicle credits for New York State. I've also published reports with the Electric Power Research Institute, NYSERDA, and the U.S. Department of Energy on the NYPA/TH!NK Clean Commute Program and Green Schools.

My purpose in writing this book was not just to tell people how to convert their cars to plug-in hybrid electric cars. I also wanted to get people moving closer to the electric car as quickly as possible. The intent was to create a useful guide to get you started, to encourage you to contact additional sources in your own "try-before-you-buy" quest, to point you in the direction of the people who have already done it (i.e., electric vehicle associations, consultants, builders, suppliers, and integrators), to familiarize you with the electric vehicle components, and finally, to go through the process of actually building/converting your own electric vehicle. My intention was never to make a final statement, but only to whet your appetite for electric vehicle possibilities. I hope you have as much fun reading about the issues, sources, parts, and building process as I have had writing about them.

Seth Leitman

Acknowledgments

While I will thank all the people who helped with this book, I want first to thank my family. Its members have watched my involvement in electric cars expand over the years, and with global warming and green becoming the new black, they really appreciate what I am doing. More important, I am in this movement so that my children can have a better earth to live in, since climate change and global warming are a reality.

I dedicate this book to my beautiful wife, Jessica, and my beautiful sons, Tyler and Cameron (to whom I am hoping this book helps to bring a better world as they grow up). They have seen me through everything and encourage me now while I write my books. It's awesome! While sometimes things have been difficult for a person starting a company, Jessica has always supported me, loved me, wished for my happiness, and was a rock during the tough times. Tyler and Cameron are two great boys who inspire me with their happiness, love, and intellect. As I see them grow up in the world we live in today, I am glad that this book can do its part for their future.

There are so many other people to thank.

I'd like to thank Felix Kramer of CalCars.org, Remy Chevalier of Electrifying Times, Josh Dorfman of The Lazy Environmentalist, and Chelsea Sexton of Plug In America. All are working to make sure plug-in hybrid electric cars become a reality. They do it in a bi-partisan manner and they do it for all the right reasons. Felix Kramer once told me that he always tells people that the best type of car is an electric car.

I also want to thank Steve Clunn. He assisted with the book and is an electric vehicle conversion specialist. He owns a company in Florida called Grass Roots EV, and we have been talking about car conversions for the past three years. I have always supported him and believed he was doing great things. Now he has so many cars to convert he has a waiting list. That is great for him and a great sign for the conversion industry.

Of course, I *must* thank Judy Bass from McGraw-Hill. She is such a sweet, loving person who believed in me. I have always believed that people come into your life for a reason. Her reason was to bless me with the opportunity to write this book and give this industry the jolt it needs.

CHAPTER 1

Why Plug-In Hybrid Electric Cars Can Happen Now!

Would you like a vehicle that gets 100 miles per gallon (Figure 1-1)? A plug-in hybrid electric vehicle will do just that.

What Is a Plug-In Hybrid Electric Vehicle?

What is a plug-in hybrid electric vehicle? Why should I take a hybrid electric car and convert it to a plug-in hybrid?

The best way I can put it is to say that a plug-in hybrid is cleaner and more energy-efficient than a hybrid electric car. A plug-in hybrid can be a gas car with electric batteries that have a range of 20, 30, 40, 50, 60, or 70 miles; or it can be a hybrid electric car that has a purely zero-emission vehicle (ZEV) range of 20, 30, 40, 50, 60, or 70 miles.

Why Should You Convert Your Car to a Plug-In Hybrid Electric Vehicle?

If you use your car for commuting to work or driving around town, a plug-in hybrid acts as an electric car all the time you are driving. How important is that? Well, let's put it this way. I am an electric vehicle purist at heart, and to transform the automobile market, we need more electric and fewer gasoline-powered cars.

You should convert your car simply because a plug-in hybrid electric car is one of the cleanest, most efficient, and most cost-effective forms of transportation around—*and* it is really fun to drive.

Plug-in hybrid electric vehicles (PHEVs) combine the benefits of pure electric vehicles and those of hybrid electric vehicles. Like pure electric vehicles, they plug into the electric grid and can be powered by the stored electricity alone. Like hybrid electric vehicles, they have engines that enable them to have a greater driving range and that can recharge the battery.

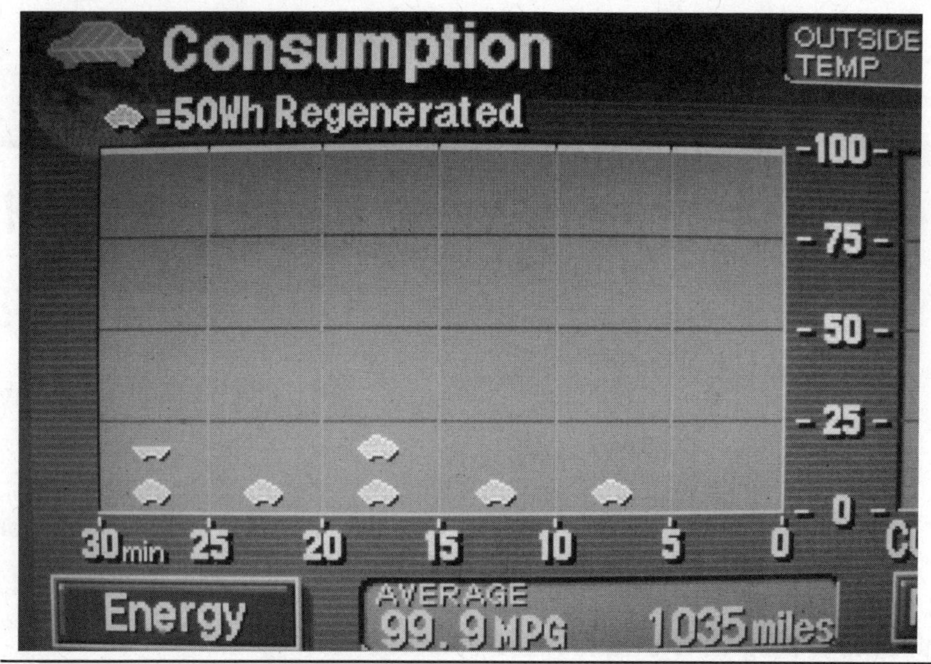

FIGURE 1-1 Yes, 99.9 or 100 miles per gallon!

The cost of the electricity needed to power plug-in hybrids for all-electric operation has been estimated at less than one-quarter of the cost of gasoline.[1] Compared to conventional vehicles, PHEVs can reduce air pollution, dependence on petroleum, and the greenhouse gas emissions that contribute to global warming. Plug-in hybrids use no fossil fuel during their all-electric range if their batteries are charged from nuclear and renewable energy sources. Other benefits include improved national energy security, fewer fill-ups at the filling station, the convenience of recharging at home, opportunities to provide emergency backup power to the home, and vehicle-to-grid applications.[2]

I remember when I worked for the State of New York at the New York Power Authority and my boss at the time, Bart Chezar, retired. Before leaving, he told me that the next step toward full electric vehicles would be the plug-in hybrid and that I should never lose sight of that. I never did, and that is the message of this book.

But wait, there's more:

- If your driving is mostly local, you'll almost never need to gas up.
- Lifetime service costs are lower for a vehicle that is mainly electric.
- A PHEV can provide power to an entire home in the event of an outage; a fleet of PHEVs could power critical systems during emergencies.

FIGURE 1-2 Hybrids Plus Prius PHEV battery packs. Photo courtesy of Davide Andrea. Source: Wikipedia.

Understanding the Plug-In Hybrid

To really appreciate the plug-in hybrid electric vehicle, it is best to examine the components of the

- Internal combustion engine
- Electric car
- Hybrid electric car
- Plug-in hybrid electric vehicle

The differences are really a study in contrasts.

Internal Combustion Engine

Mankind has been fascinated with the internal combustion engine vehicle, and this is simply an enigma. The internal combustion engine is a device that inherently tries to destroy itself: numerous explosions drive its pistons up and down to turn a shaft. A shaft rotating at 6,000 revolutions per minute produces 100 explosions every second. These explosions, in turn, require a massive vessel to contain them—typically a cast iron cylinder block. Additional systems are also necessary:

- A cooling system to keep the temperature within a safe operating range
- An exhaust system to remove the heated exhaust products safely
- An ignition system to initiate the combustion at the right moment
- A fueling system to introduce the proper mixture of air and gas for combustion
- A lubricating system to reduce wear on high-temperature, rapidly moving parts
- A starting system to get the whole cycle going

It's complicated to keep all these systems working together. This complexity means that more things can go wrong (more frequent repairs and higher repair costs). Figure 1-3 summarizes the internal combustion engine vehicle systems.[3]

Unfortunately, the destruction caused by the internal combustion engine vehicle doesn't stop with itself. The internal combustion engine is a variant of the generic combustion process. To light a match, you use oxygen (O_2) from the air to burn a carbon-based fuel (a wood or cardboard matchstick), generate carbon dioxide (CO_2), emit toxic waste gases (you can see the smoke and perhaps smell the sulfur), and leave a solid waste (the burnt matchstick). With a match, the volume of air around you is far greater than that consumed by the match; thus, air currents soon dissipate the smoke and smell, and you toss the matchstick.

Today's internal combustion engine is more evolved than ever. However, it still uses a carbon-based combustion process that creates heat and pollution. Everything about the internal combustion engine is toxic, and it is one of the least efficient mechanical devices on the planet. Unlike the lighting of a single match, the use of hundreds of millions (soon to be billions) of vehicles with internal combustion engines threatens to destroy all life on our earth. You'll read about the environmental problems caused by internal combustion engine vehicles in Chapter 2.

While an internal combustion engine has hundreds of moving parts, an electric motor has only one. That's one of the main reasons why electric cars are so efficient. To make the car, pickup, or van that you are driving now into an electric vehicle, all you need to do is take out the internal combustion engine, along with all related ignition, cooling, fueling, and exhaust system parts, and add an electric motor, batteries, and a controller. Hey, it doesn't get any simpler than this!

Figure 1-4 shows all there is to it. Batteries and a charger are your "fueling" system, an electric motor and controller are your "electrical" system, and the "drive" system is just as it was before (although today's advanced electric vehicle designs don't even need the transmission and drive shaft).

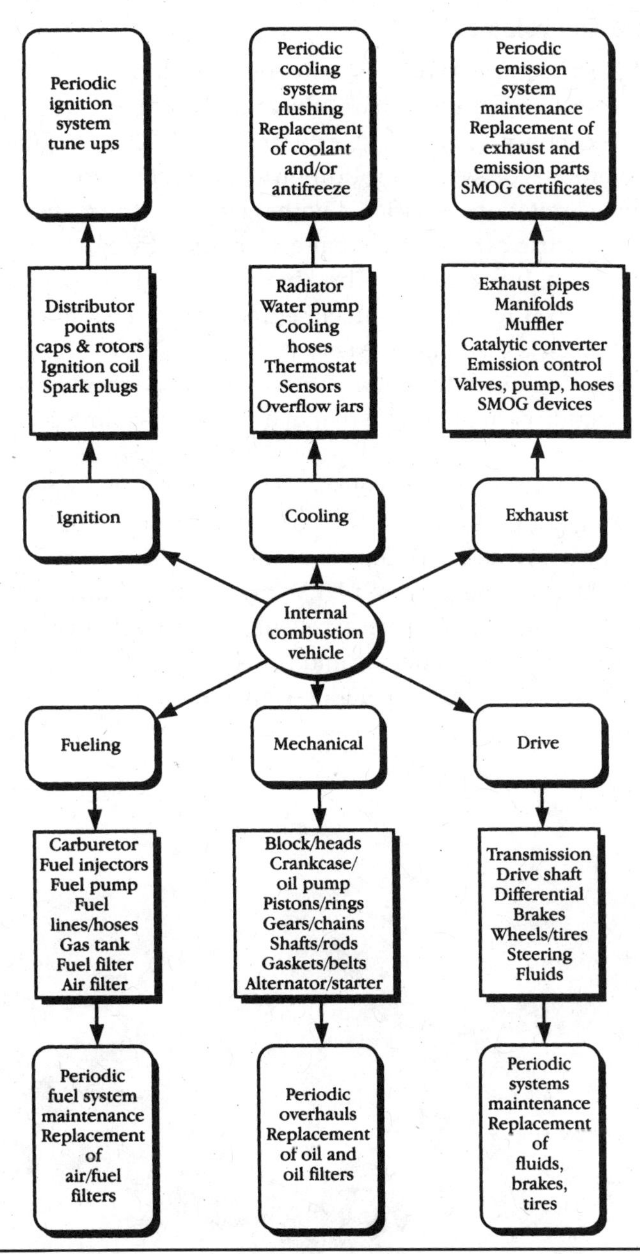

FIGURE 1-3 Internal combustion engine vehicle systems.

Hybrid Electric Vehicles

A *hybrid electric vehicle* (HEV) is a hybrid vehicle that combines a conventional propulsion system with a rechargeable energy storage system (RESS) to achieve better fuel economy than a regular internal combustion engine car (Figure 1-4). It includes a propulsion system in addition to the electric motors, so it is not hampered by the range limitations associated with the battery the way that an electric car is.

Hybrid electric cars keep the battery charged by capturing the kinetic energy generated through regenerative braking or through an electric motor/generator regulated by a controller. This can either recharge the batteries or add power to the motor that propels the vehicle. Many HEVs will reduce emissions while the car is idling by shutting down the engine and restarting it when it needs the energy. Also, an HEV's engine is smaller than that of a regular gasoline-powered car, providing more efficiency.

Right now, hybrid electric vehicles are available at almost every price point. If you are trying to figure out how to fit one of them into your life, you can buy a new HEV, such as a Toyota Prius, for a bit more than $20,000 or a used one starting at about $10,000. Gasoline-powered HEVs are not the ultimate answer to our energy problems, but they do provide an excellent platform for developing EV components such as electric motors, batteries, and transmissions. They also use much less gas than their internal combustion engine–only brethren.

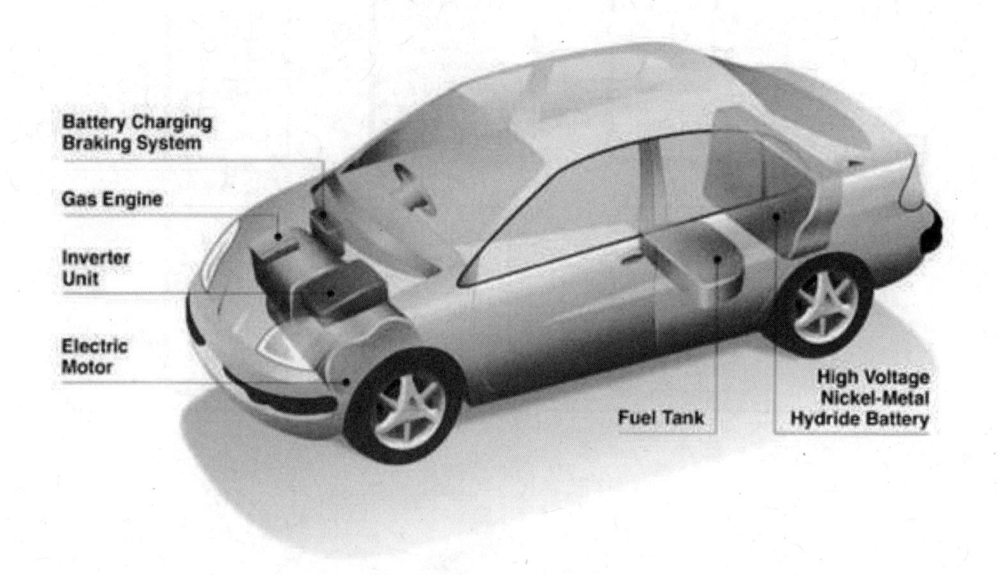

FIGURE 1-4 A diagram of a hybrid electric car. Source: *BusinessWeek*. Data: Toyota Motor Corp.; illustration by Joe Calviello.

As you can see in Figure 1-5, sales of hybrids on a monthly basis from January 2004 to February 2007 show the Prius as the consistent leader, followed by the Honda Civic Hybrid and then by the Toyota Camry Hybrid.

Electric Vehicles

A simple diagram of an electric vehicle looks like a simple diagram of a portable electric shaver: a battery, a motor, and a controller or switch that adjusts the flow of electricity to the motor to control its speed (see Figure 1-6). That's it. Nothing comes

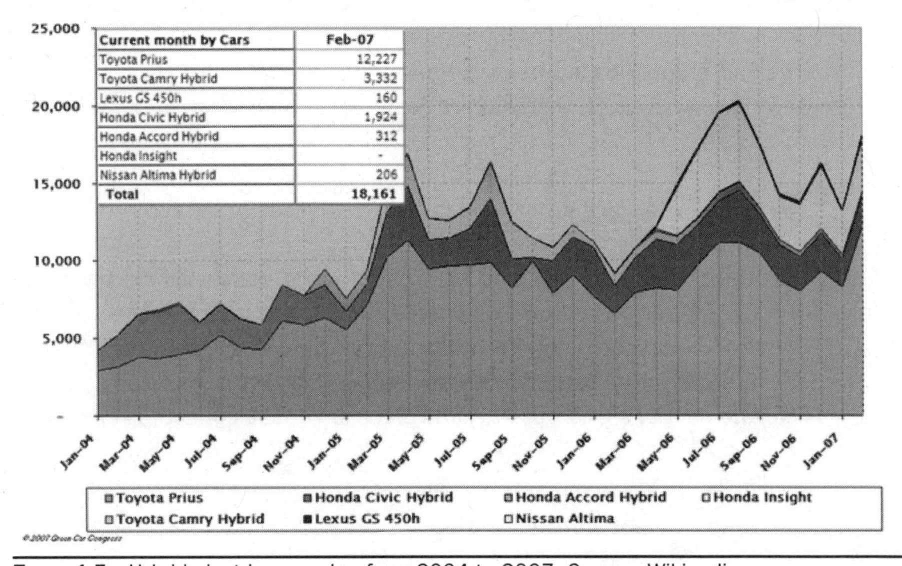

FIGURE 1-5 Hybrid electric car sales from 2004 to 2007. Source: Wikipedia.

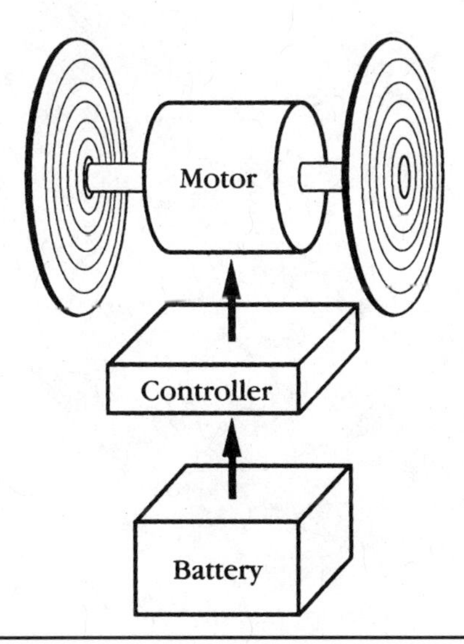

FIGURE 1-6 Electric vehicle systems.

out of your electric shaver, and nothing comes out of your electric car. EVs are simple (and therefore highly reliable), have lifetimes measured in millions of miles, need no periodic maintenance (filters and other such items), and cost significantly less per mile to operate. They are highly flexible as well, using electric energy, which is readily available anywhere, as input fuel.

In addition to all these benefits, if you buy, build, or convert your electric vehicle using the chassis from an internal combustion engine vehicle, as suggested in this book, you are performing a double service for the environment: you remove one polluting car from the road, and you put one nonpolluting electric vehicle to service.

You've had a quick tour and a side-by-side comparison of electric vehicles and internal combustion engine vehicles. Now let's take a closer look at electric vehicles.

What Is an Electric Vehicle?

An electric vehicle consists of a battery that provides energy, an electric motor that drives the wheels, and a controller that regulates the energy flow to the motor. Figure 1-7 shows all there is to it—but don't be fooled by its simplicity. Scientists, engineers, and inventors down through the ages have always said, "In simplicity there is elegance." Let's find out why the electric vehicle concept is elegant.

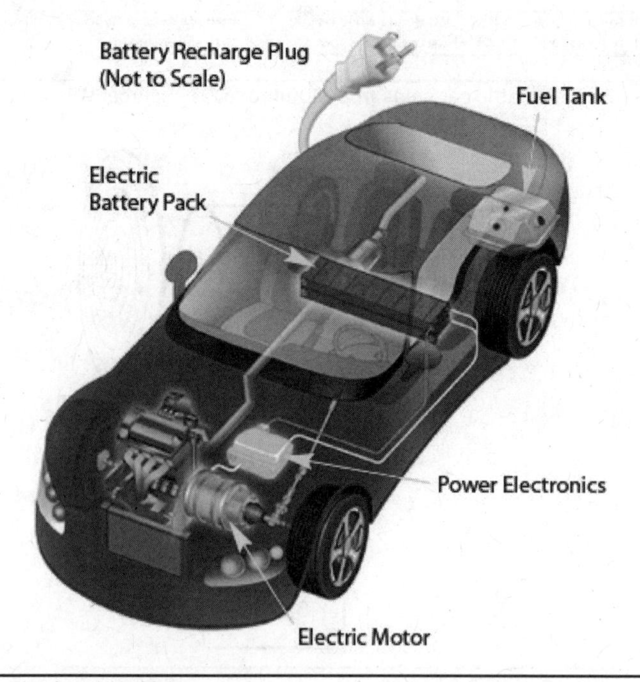

FIGURE 1-7 A basic electric vehicle.

Electric Motors

Electric motors can be found in so many sizes and places, and have so many varied uses, that we tend to take them for granted. Universal in application, they can be as big as a house or smaller than your fingernail, and they can be powered by any source of electricity. In fact, they are so reliable, quiet, and inexpensive that we tend to overlook just how pervasive and influential they are in virtually every civilized person's life.

Each of us encounters dozens, if not hundreds, of electric motors daily without even thinking about them. The alarm clock wakes you; you turn on the television for the news; you remove coffee beans from the refrigerator and put the coffee beans in a grinder; in the bathroom, you use an electric shaver, electric toothbrush, or hair dryer; making breakfast might be assisted by your electric juicer, blender, or food mixer; you might clean your home with your vacuum cleaner or clean your clothes with your washer and dryer. Next, you use your automobile or a subway, bus, or light rail transit to ride to work, where you might go through an automatic gate or door or take an elevator or escalator to your floor; at home or at work, you sit down at your computer, use the Internet or your e-mail, cell phone, or BlackBerry, and use your fax or copier after you adjust the fan, heater, or air conditioner. Back at home in the evening, you might use an electric garage door opener, program your TiVo, or use an electric power tool on a project. You get the picture.

Why are electric motors ubiquitous? In a word, *convenience*. Electric motors do work so that you don't have to. Whether it's pulling, pushing, lifting, stirring, or oscillating, the electric motor converts electric energy into motion, which is further adapted to do useful work.

What is the secret of the electric motor's widespread use? Its reliability. This is because of its simplicity. Regardless of type, all electric motors have only two basic components: a rotor (the moving part) and a stator (the stationary part). That's right—an electric motor has only one moving part. If you design, manufacture, and use an electric motor correctly, it is virtually impervious to failure and indestructible when in use.

In internal combustion automobiles, in addition to your all-important electric starter motor, you typically find electric motors in the passenger compartment heating/cooling system, radiator fan, windshield wipers, electric seats, windows, door locks, trunk latch, outside rearview mirrors, outside radio antenna, and many other places.

Batteries

No matter where you go, you cannot get away from batteries, either. They're in your pocket tape recorder, portable radio, telephone, cell phone, laptop computer, portable power tool, appliance, game, flashlight, camera, and many other devices. Batteries come in two distinct flavors: rechargeable and nonrechargeable. Like motors, they come in all sorts of sizes, shapes, weights, and capacities. Unlike motors, they have no moving parts. With nonrechargeable batteries, you simply dispose of them when they are out of juice; with rechargeable batteries, you connect them to a recharger or source of electric power to bring them back up to capacity. There are different types of batteries. Among rechargeable batteries, there are lead-acid, nickel metal hydride, and lithium ion batteries (as some examples); these can be used in your car to manage the recharging process invisibly via an under-the-hood generator or alternator that recharges the battery while you're driving.

Why are batteries ubiquitous? In a word: *convenience*. In an automobile powered by a conventional internal combustion engine, the battery, in conjunction with the starter motor, serves the all-important function of starting the automobile. In fact, it was the battery and electric starter motor combination, first introduced in the early 1920s and changed very little since then, that put the internal combustion engine car on the map—it made cars easy to start and easy to use for anyone, anywhere.

Another great thing about the promise of electric cars is lithium-ion battery technology. It is moving into the marketplace very fast and dropping in price. Over the next few years, we can expect further drops in price, making PHEV conversions more affordable. Soon enough, lithium-ion batteries will be standard in any conversion kit.

Rechargeable lead-acid automotive batteries perform their job very reliably over a wide range of temperature extremes and, if kept properly charged, will maintain their efficiency and deliver stable output characteristics over a relatively long period of time—several years. A lead-acid automotive battery is unlikely to fail unless you shock it, drop it, discharge it completely, or allow a cell to go dry. The only maintenance required with lead-acid batteries is checking each cell's electrolyte level and refilling them with water periodically. Newer, sealed batteries require no maintenance at all.[4]

Controllers

Controllers have become much more intelligent. The same technology that reduced computers from room-sized to desk-sized allows you to exercise precise control over an electric motor. Regardless of the voltage source, current needs, or motor type, today's controllers—built with reliable solid-state electronic components—can be designed to meet virtually any need and can easily be made compact to fit conveniently under the hood of your car.

Why are electric vehicles elegant? When you join an electric motor, battery, and controller together, you get an electric vehicle that is both reliable and convenient. Perhaps the best analogy is that when you "go EV," you can drive your entire car from an oversized electric starter motor, a more powerful set of rechargeable batteries, and a very sophisticated starter switch. But it's only going to get better.

Back in the early 1990s, when the first edition of this book was published, electric vehicles resembled your battery-operated electric shaver, portable power tool, or kitchen appliance. Today's and tomorrow's electric vehicles more closely resemble your portable laptop computer in terms of both sophistication and capabilities.

FIGURE 1-8 Honda Insight hybrid electric car (the most fuel-efficient hybrid). Source: Wikipedia.

FIGURE 1-9 Toyota Prius (the leading hybrid). Source: Wikipedia, IFCAR, All Rights Released.

Convert That Hybrid!

Now that hybrid electric cars are part of the future, why not go further—why not make that hybrid an electric car? Why not turn that car into a zero-emission vehicle? We can and we should.

It is not difficult to convert hybrid electric cars to plug-in hybrids. There are many reasons that should convince you of the need to go to a plug-in hybrid:

- The cost of a gallon of gas
- Higher asthma rates
- Our need to reduce our reliance on imported oils
- The prospect of owning a car that is cost-effective, fun, and longer-lasting than most cars on the road today

Once again, how about the fact that you can convert a plug-in hybrid electric vehicle *today*? Right now! And doing so will cost you less than some new cars on the market.

Reduce Our Reliance on Oil and Clean the Environment at the Same Time!

July 2007 EPRI-NRDC Definitive Study: PHEVs Will Reduce Emissions if Broadly Adopted

In July 2007, the Electric Power Research Institute (EPRI) and the National Resources Defense Council (NRDC) did one report that included multiple studies on plug-in hybrids.[5] The report stated that scientists have confirmed that, unlike gasoline cars, plug-ins will get cleaner as they get older—because our power grid is getting cleaner.

For people who are looking for the most effective way to end our addiction to oil, PHEVs make sense because carmakers can build them now, with today's technology and using today's infrastructure. The study showed that with the increase in the number of PHEVs on the road and the evolving characteristics of the power grid (in terms of capacity and carbon intensity), PHEVs will vastly reduce greenhouse gases over the next 40 years. The second study showed that increased PHEV use will reduce greenhouse gases over the next 20 years; even if, in the worst-case scenario, we still use lots of coal, nationwide air quality in terms of other emissions will also improve.

Three more points: both reports match up well with previous studies. They reinforce the Pacific National Lab's January 2007 findings that we won't have to build new power plants for cars that charge at night. And we're gratified that General Motors recognizes this study as validation of its decision to evolve toward the electrification of transportation.

California Air Resource Board/ZEV States Report

California, New York, Massachusetts, and some other states have had zero-emission-vehicle programs since the early 1990s because battery electric vehicles in those states, taking into account power plants, are far cleaner than gasoline cars in reducing urban air pollution and smog. The issue keeps being raised, although studies are conclusive.

The "well-to-wheel" emissions of electric cars are lower than those from gasoline internal combustion vehicles. A California Air Resources Board (CARB) study showed that *battery electric vehicles emit at least 67 percent less greenhouse gases than gasoline cars*—even more assuming the power is generated with renewable energy sources. A PHEV with only a 20-mile all-electric range emits 62 percent less.

U.S. DOE Argonne National Lab

Two government studies have found that PHEVs would result in large reductions even on the national grid (which is 50 percent coal). The GREET 1.6 emission models done in 2001 by the DOE's Argonne National Lab estimates that hybrids reduce greenhouse gases by 22 percent and plug-in hybrids by 36 percent (see Figure 1-10). An Argonne researcher reached consensus with researchers from other national labs, universities, the Air Resources Board, automakers, utilities, and Arthur D. Little to estimate in July 2002 that PHEVs using nighttime power reduce greenhouse gases by 46 to 61 percent.

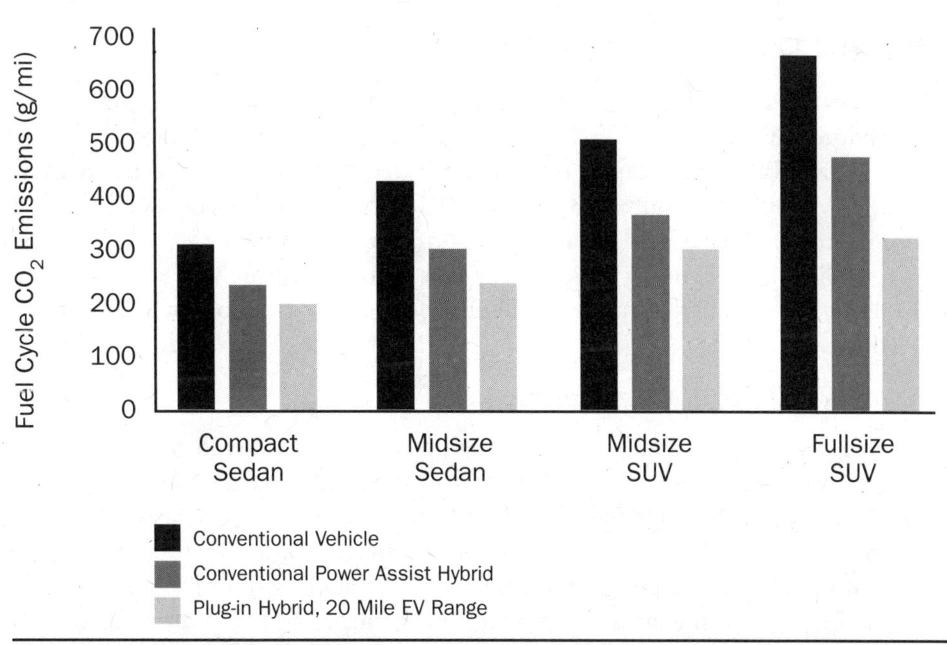

FIGURE 1-10 EPRI and NRDC emissions estimates from PHEVs. Source: http://www.calcars.org/vehicles.html.

FIGURE 1-11 A plug-in hybrid extension cord would look like this. Simple, isn't it? Source: http://www.metaefficient.com/images/plug-in-hybrid-car-phev.jpg.

PHEVs Are Cheaper to Run and Maintain

A 2006 research estimate in California found that the operating costs of plug-ins charged at night were equivalent to $0.75 per U.S. gallon (3.8 L) of gasoline.[6]

The EPRI-NRDC report further concluded that the cost of electricity for a Prius PHEV would be about $0.03 per mile ($0.019 per kilometer), based on 0.26 kWh/mi (129 mpg) and a cost of electricity of $0.10 per kilowatt hour.[7]

During 2008, many government and industry researchers have been focusing on determining what range for all-electric operation is economically optimum for the design.[8]

20 to 25 Cents per Gallon?!

Using the average U.S. electricity rate of 9 cents per kilowatt-hour (kWh), *30 miles of electric driving will cost 81 cents*. If we optimistically assume that the average fuel economy in the United States is 25 miles per gallon, at $3.00 a gasoline, this equates to *75 cents a gallon* for the equivalent electricity. Compared to a regular hybrid's real-world 45 miles per gallon, it's effectively $1.20 a gallon.

PHEVs are meant to be plugged in at night. In many areas of the country, overnight power is available at a lower cost. As PHEVs start to enter the marketplace, we'll see increasing support from electric utilities; they'll offer reduced nighttime rates to provide incentives for off-peak charging. In some areas where wind and hydropower are wasted at night, the rate can be as low as 2 to 3 cents per kWh. That's 20 to 25 cents a gallon.

Purchase Costs

Calcars and others are charging $15,000 for a conversion. However, a conversion by building your own will cost much less and will increase the payback.

People routinely pay more for such options as sunroofs, automatic transmissions, V8 engines, and leather seats. These are "features"—and no one asks about the payback. A J.D. Power survey shows that buyers will pay more for cars with the "environmental feature." How much more? The high demand for the Honda Civic Hybrid tells us that it's at least $3,000.

The Bottom Line

A 2003 EPRI battery study shows that mass-produced PHEVs *have already reached life-cycle cost parity with gas-powered vehicles*—using gas prices from three years ago!

This means that the more maintenance-free electrical systems of PHEVs offset the initial higher cost of batteries.

Payback?

The costs and benefits of cars extend far beyond an individual driver to society as a whole. But when people talk about payback, they refer only to the net dollars to the driver. Because this question never comes up when people pay a premium for features like leather seats, we point out that millions of people want the "environmental feature" (see J.D. Power and Associates' 2004 report). *Car and Driver*'s Patrick Bedard writes amusingly but tellingly about this issue.

Despite this, a 2003 EPRI study, *assuming only $2 a gallon gas*, zero buying incentives, and a PHEV premium of $3,000 to $5,000 more than standard hybrids, shows that the total lifetime cost of ownership for a PHEV will be lower than that for any other vehicle type—so the payback will be there.[9]

Safety First

On another safety issue, while electric vehicles do not emit noise pollution, there has been concern about hybrid vehicles being unsafe for visually impaired pedestrians because their engines don't make noise.

The Baltimore-based National Federation of the Blind presented written testimony to the U.S. Congress asking that a minimum sound standard for hybrids be included in the emissions regulations. The president of the group, Marc Maurer, stated that he's not interested in returning to gas-guzzling vehicles; the group just wants fuel-efficient hybrids to have some type of warning noise. "'I don't want to pick that way of going, but I don't want to get run over by a quiet car, either," Maurer said.

Manufacturers are aware of the problem but have made no pledges yet. Toyota is studying the issue internally, said Bill Kwong, a spokesman for Toyota Motor Sales USA.

"One of the many benefits of the Prius, besides excellent fuel economy and low emissions, is quiet performance. Not only does it not pollute the air, it doesn't create noise pollution," Kwong said. "We are studying the issue and trying to find that delicate balance."

The Association of International Auto Manufacturers Inc., a trade group, is also studying the problem, along with a committee established by the Society of Automotive Engineers. The groups are considering "the possibility of setting a minimum noise level standard for hybrid vehicles," said Mike Camissa, the safety director for the manufacturers' association.[10]

You Can Do It!

What is happening with plug-in hybrid electric vehicles is merely the beginning. As battery technology improves, PHEVs will get faster, have a longer range, and be even more efficient. All the available technology has just about been squeezed out of internal combustion engine vehicles, and they are going to be even more environmentally squeezed in the future. This will hit each buyer right in the pocketbook—incremental gains will not come inexpensively. Internal combustion engines are nearly at the end of their technological lifetime.

Electric vehicles have also been around for more than 100 years, so making a hybrid car and PHEV will greatly improve their range and performance. When you do it yourself, any choice you wish to make for more speed, acceleration, or range is readily accommodated. Just do it.[11]

CHAPTER 2

PHEVs Save the Environment and Energy

The needs of the many outweigh the needs of the few.
Mr. Spock *(from Star Trek: The Wrath of Khan)*

Besides the fact that consumers have been consistently interested in the electric car (as indicated by popular reporting by car companies), there is a newly sparked (no pun intended) excitement about both plug-in hybrid electric vehicles (which are more electric car than hybrid) and electric cars (Tesla, TH!NK City, RAV4, EV1). This means only great things for the planet—specifically, zero tailpipe emissions and better air quality in our major metropolitan areas. And with this comes a significant reduction in overall energy use.

How Do PHEVs Save the Environment?

Because they are zero-emission vehicles (ZEVs). PHEVs do not emit toxic compounds into our atmosphere. In contrast, everything that goes into and comes out of an internal combustion vehicle is toxic, and it's among the least efficient mechanical devices on the planet. (The toxic emissions of the power plants used to supply the electricity used by PHEVs are held to a lower level [higher standard] than those from gasoline-powered vehicles.)

Far worse than the inefficient and self-destructive operating nature of the internal combustion engine is the legacy of environmental problems (summarized in Figure 2-2) created by the hundreds of millions or billions of vehicles that these engines power:

- Dependence on foreign oil (an environmental and national security risk)
- Greenhouse effect (atmospheric heating)
- Toxic air pollution
- Wasted heat generated by these engines' inefficiency

FIGURE 2-1 How about $0.75 a gallon? Source: CalCars.org.

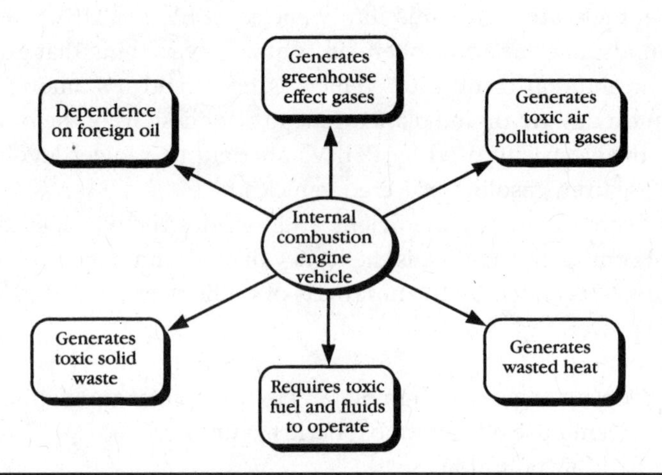

FIGURE 2-2 Internal combustion engine vehicles create many problems.

Save the Environment and Save Some Money Too!

Because PHEVs use less energy than gasoline-powered vehicles, their effect on the environment is much less than that of vehicles powered by fossil fuels. Because PHEVs are more efficient then petroleum-powered vehicles, they cost less to run.

While I believe it is unquestionable that EVs will be the de facto transportation mode of choice in the years to come, PHEVs are the next step in this direction that the consumer marketplace can accept at this time, since there are hybrid electric vehicles on the road that need only to be electrified.

Figure 2-3 shows the reasons why. The EPRI report discussed in Chapter 1 talks about emission reductions resulting from PHEV use. Figure 2-3 shows the serious CO_2 emission reductions that PHEVs have produced to date.

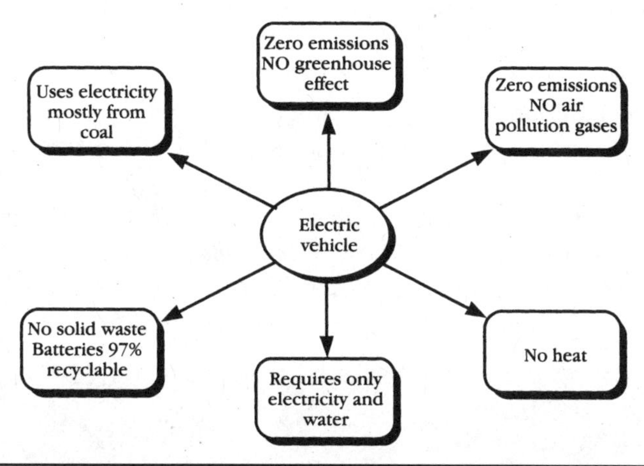

FIGURE 2-3 PHEV emission reductions found by the EPRI study. Source: http://www.calcars.org/vehicles.html.

Fuel-Efficient Vehicles

Car companies are now responding by developing or providing more fuel-efficient cars. The market wants them, and the car companies that are doing well (Toyota and Honda) are producing fuel-efficient and hybrid cars. Now, Ford, GM, and the other car companies are either producing or developing cars of these types. This is to be expected in business.

Much has been said about creating a new type of car that can get 35 mpg. In other countries, where fuel is very expensive, such cars already exist. These cars are much smaller and have smaller engines. What has not been said is the great extent to which the vehicle's driver controls its fuel efficiency. In countries where fuel is expensive, drivers tend to drive at slower speeds. Driving twice as fast requires four times as much energy to overcome aerodynamic losses. Driving at 100 mph rather than 50 mph increases the rate at which fuel or electric energy is used almost by a factor of eight. (Since you get there in half the time, the total energy used is increased by a factor of four.) More than anything else, it is the driver's right foot that controls mpg or mi/kWh for a particular vehicle. *Even if you do not have an electric car, plan on converting a car, or buy a hybrid electric car, one thing to take away from this book is to drive more efficiently to reduce your carbon footprint.*

So Who's to Blame?

No one and everyone. The line from *Dr. Zhivago* that says, "One Russian ripping off wood from a fence to provide heat for his family in winter is pathetic, one million Russians doing the same is disaster," applies equally well to all of us and our internal combustion engine vehicles. Applied collectively, the legacy of the internal combustion engine is the greenhouse effect, dependence on foreign oil, and pollution. Let's succinctly define the problem and its solution.

U.S. Transportation Depends on Oil

Although small amounts of natural gas and electricity are used, the U.S. transportation sector is almost entirely dependent on oil. A brief look at a few charts (Figures 2-4 and 2-5) will demonstrate the facts. It doesn't take a rocket scientist to figure out that this situation is both a strategic and an economic problem for us.

As mentioned in the second edition of *Build Your Own Electric Vehicle*, 40 percent of our energy comes from petroleum, 23 percent from coal, and 23 percent from natural gas. The remaining 14 percent comes from nuclear power, hydroelectric

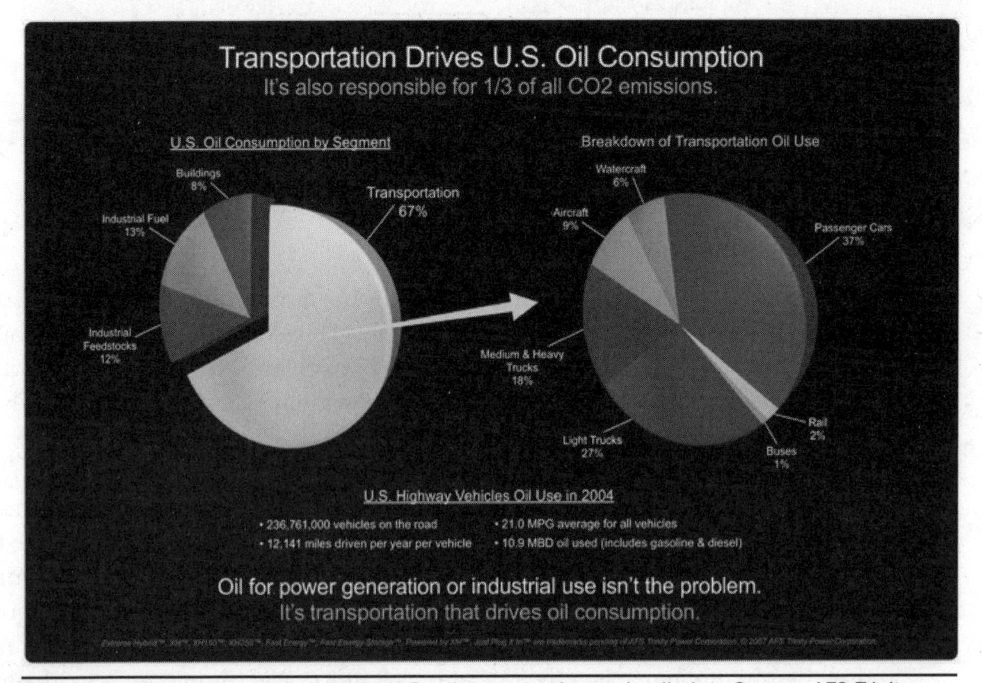

FIGURE 2-4 How transportation drives U.S. oil consumption and pollution. Source: AFS Trinity.

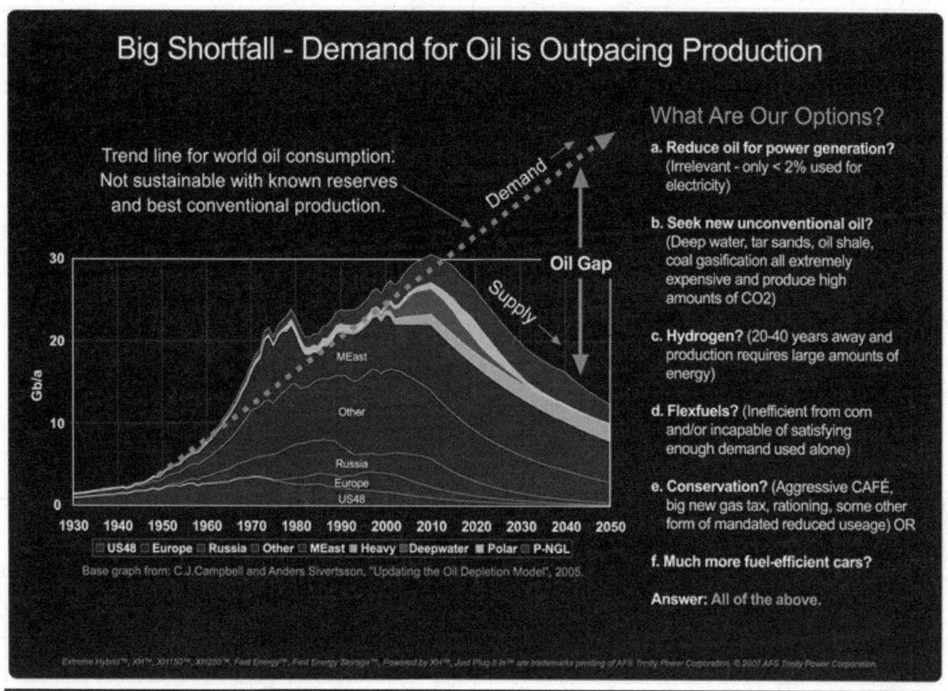

FIGURE 2-5 Oil is a finite resource. Source: AFS Trinity.

power, and renewable sources. As Bob Brant stated, "Our entire economy is obviously dependent on oil."

The United States consumes 20.8 million barrels of petroleum a day, of which 9 million barrels are in the form of gasoline. Automobiles are the single largest consumer of oil, consuming 40 percent, and are also the source of 20 percent of the nation's greenhouse gas emissions.[1]

The United States has about 22 billion barrels of reserves and consumes about 7.6 billion barrels per year.[2]

Problems associated with the U.S. oil supply include volatile oil prices, increasing world and domestic demand, and falling domestic production.

While it's our own fault for letting this happen, the Organization of Petroleum Exporting Countries (OPEC) price hikes have had a disastrous impact on our economy, our transportation system, and our standard of living. The Arab oil crisis of 1973 and subsequent crises were not pleasant experiences. After each such crisis, the United States has vowed to become less dependent on foreign oil producers—yet exactly the opposite has happened.

Increasing Long-Term Oil Costs

There is a fixed amount of oil/petroleum reserves in the ground around the world, and there isn't going to be any more. We're going to run out of oil at some point. Before that happens, it's going to get very expensive.

How did we get into this situation? We put ourselves into it. Figure 2-4 shows that in 2005, it was estimated that the price of a barrel of oil would reach $100. We have already seen prices above that level. No one can accurately predict what fuel prices will be this summer or next year, or whether there will be a shortage or an abundance of supplies. Everyone agrees that this is a bad situation. We need to take real steps to correct the problem. Since none of us has the luxury of ceasing to go to work, many creative "work-around" solutions would come forward if gasoline were to remain at $4 a gallon in the United States or the price for a barrel of oil were to hold at more than $100—for example, PHEVs.

How PHEVs Can Help

Maintaining stringent controls on the emission of toxic air pollutants and conforming to increasingly higher mandated corporate average fuel economy levels puts an enormous burden on internal combustion engine vehicle technology and on your pocketbook. Automotive manufacturers have to work their technical staffs overtime to accomplish these feats, and the costs will be passed on to the new buyer.

To quote Wang, DeLuchi, and Sperling,[3] who studied the subject extensively:

> The unequivocal conclusion of this paper is that in California and the United States the substitution of PHEVs for gasoline-powered vehicles will dramatically reduce carbon dioxide, hydrocarbons and to a lesser extent, nitrogen oxide emissions.

While the reduction in petroleum consumption for a particular PHEV depends on how the vehicle is designed and used, Wang estimates that a PHEV may consume nearly 60 percent less gasoline than a conventional vehicle, and almost 30 percent less gasoline than a nonpluggable HEV2. Since PHEVs use less gasoline, they also emit less greenhouse gases (and potentially fewer criteria pollutants) from the tailpipe. In fact, some PHEV designs with robust all-electric driving capability, i.e., the ability to accelerate and cruise all-electrically in real-world driving, would have no tailpipe emissions at all until the onboard electric energy storage is discharged to its design minimum.[4]

According to the EPRI report:

> However, tailpipe emissions are not the only consideration. To fully assess the environmental impact of PHEVs, lifecycle emissions (including the emissions that result from upstream electricity generation) must be evaluated. Not surprisingly, the lifecycle emissions benefits of PHEVs are dependent on the fuel sources for the electricity used to recharge them. Recharging PHEVs with renewable electricity dramatically reduces greenhouse gas and criteria pollutant emissions, while using electricity from coal yields less impressive results. Given the complexity of such analyses, a variety of conclusions have been reached; the variation depends on assumptions, inputs, and nomenclature. In general though, results from such studies indicate that PHEVs are likely to reduce greenhouse gas emissions associated with transportation and to reduce exposure to other regulated emissions for most Americans.

> Wang estimated that a gasoline-powered PHEV using "average" electricity from the U.S. grid would emit 37 percent fewer greenhouse gases than a conventional vehicle, but would increase emissions of nitrogen oxides (NOX) six percent, particulate matter (PM10) 3.5 percent, and sulfur oxides (SOX) 62 percent. Kliesch and Langer estimate large regional differences in PHEV emissions benefits.[5]

The comments in the greenhouse gas section of the EPRI report on power plant emissions associated with the production of electricity for electric vehicle use are equally applicable to air pollution. In addition, shifting the burden of producing electricity for electric vehicles to coal-powered generating plants (often located out of state) has these effects:

- It focuses pollution control on smokestack "scrubbers" and other mandated controls on stationary sites that are far more controllable than the tailpipes of vehicles with internal combustion engines.
- It shifts automotive emissions from congested, populated areas to the remote, less populated areas where many coal-fired power plants are located.
- It shifts pollution associated with automotive emissions to the nighttime (when most PHEVs will be recharged), when fewer people are likely to be exposed to them and when emissions in the atmosphere are less likely to react with sunlight to produce smog and other by-product pollutants.

Toxic Liquid and Solid Waste Pollution

Almost everything that goes into or comes out of an internal combustion engine is toxic. In addition to the internal combustion engine vehicle's greenhouse gas and toxic air pollution outputs, consider its liquid waste (fuel spills, oil, antifreeze, grease, and so on) and solid waste (oil-air-fuel filters, mufflers, catalytic converters, emission control system parts, radiators, pumps, spark plugs, and so on) by-products. This does not bode well for our environment, our landfills, or anything else—especially not when multiplied by hundreds of millions of vehicles.

Toxic Input Fluids Pollution

Remember, almost everything that goes into or comes out of the the internal combustion engine is toxic. The fuel and oil that you put into an internal combustion engine, the fuel vapors at the pump (and those associated with extracting, refining, transporting, and storing fuel), and the antifreeze you use in its cooling system are all toxic and/or carcinogenic, as a quick study of the pump and container labels will point out. On the output side, when you burn coal, oil, gas, or any other fossil fuel, you create more problems, through either the amount of carbon dioxide (CO_2) or the other types of toxic emissions produced.

Everything you pour into an internal combustion engine is toxic, but some chemicals are especially nasty. In addition to more than 200 compounds on its initial hazardous list, the Clean Air Act of 1990 amendments said:

> [The] study shall focus on those categories of emissions that pose the greatest risk to human health or about which significant uncertainties remain, including emissions of benzene, formaldehyde and 1,3 butadiene.[6]

Fouling the environment as in the *Exxon Valdez* oil spill disaster of the 1990s is one thing. Poisoning your own drinking water is another. Those enormous holes in the ground near neighborhood gas stations everywhere (as they rush to comply with federal regulations regarding acceptable levels of gasoline storage tank leakage) make this point. So does the recall of millions of bottles of Perrier drinking water when only tiny levels of benzene contamination were involved.

How can the electric vehicle help? The only substance you pour into your electric vehicle occasionally is water (preferably distilled).

Waste Heat Caused by Inefficiency

Although its present form represents its highest evolution to date, the gasoline-powered internal combustion engine is classified as being among the least efficient mechanical devices on the planet. The internal combustion engine is about 20 percent efficient. In contrast, the efficiency of an advanced DC motor generally runs between 80 and 90 percent (although sometimes lower).

In gasoline-powered vehicles, only 20 percent of the energy of combustion becomes mechanical energy; the remaining 80 percent becomes heat and is lost in the engine system. Of the 20 percent that becomes mechanical energy:

- One-third is used to overcome aerodynamic drag (the energy ends up as heat in air).
- One-third is used to overcome rolling friction (the energy ends up as heated tires).
- One-third is used to power acceleration (the energy ends up heating the brakes).

In contrast to the hundreds of moving parts in an internal combustion engine, the electric motor has just one. That's why electric motors are so efficient. Today's EV motor efficiencies are typically 90 percent or more. The same applies to today's solid-state controllers (with no moving parts), and today's lead-acid batteries come in at 75 percent or more. Combine all these and you have an electric vehicle efficiency that's far greater than anything possible with an internal combustion engine vehicle.

Electric Utilities Love PHEVs

Even the most wildly optimistic projections for electric vehicles show only a few million PHEVs in use by early in the twenty-first century. At somewhere around that level, EVs will begin making a dent in the strategic oil, greenhouse gas, and air quality problems. But until you have 10 to 20 million or more electric vehicles, you're not going to require additional electric generating capacity. This is because of the magic of *load leveling*. Load leveling means that if PHEVs are used during the day and recharged at night, they perform a great service for their local electric utility, whose demand curves almost universally look like that shown in Figure 2-6.

How electricity is generated varies widely from one geographic region to another, and even from city to city in a U.S. region. In 1991, the net fuel mix used by electric utilities was 54.87 percent coal, 21.73 percent nuclear fission, 9.75 percent hydroelectric, 9.36 percent natural gas, 3.93 percent oil, and 0.35 percent geothermal and other. Those electric utility plants that produce electricity at the lowest cost (i.e., those fueled by coal and hydro) are used to supply base-load demands, while peak demands are met by less economical generation facilities (i.e., gas and oil).[7]

When owners recharge their PHEVs in the evening hours (valley periods), they receive the benefit of an off-peak (typically lower) electric rate. Since recharging PHEVs raises the valleys and brings up its base-load demand, the electric utility is able to utilize its existing plant capacity more efficiently. This is a tremendous near-term economic benefit to our electric utilities because it represents a new market for electricity sales with no additional associated capital asset expense.

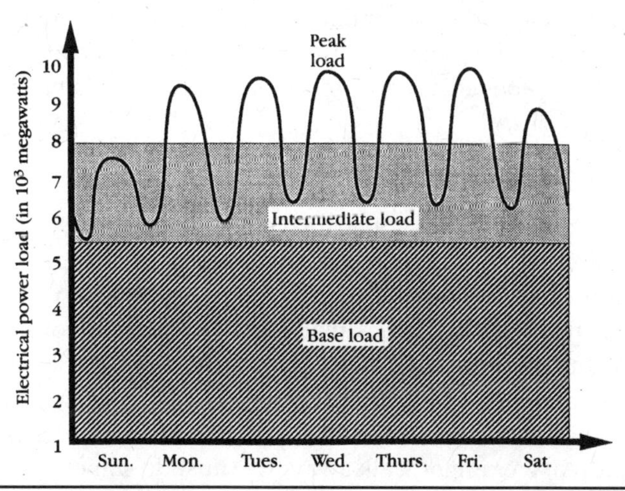

FIGURE 2-6 Weekly peak power-demand curve for a large utility operating with a weekly load factor of about 80 percent.

Chapter Summary

Electric vehicle ownership is the best first step you can take to help save the planet. But there is still more you can do. Do your homework. Write your senator or member of Congress. Voice your opinion. Get involved with the issues. But don't settle for an answer that says, "We'll study it and get back to you." Settle only for action—who is going to do what by when and why. I leave you with a restatement of the problem, a possible framework for a solution, and some additional food for thought.

The Legacy of the Internal Combustion Engine Is Environmental Problems

Internal combustion engine technology and fuel should be priced to reflect their true social cost, not just their economic cost, because of the environmental problems that they create:

- Our dependence on foreign oil security risk problem
- The greenhouse gas problem
- The air quality problems of our cities
- The toxic waste problem
- The toxic input fluids problem
- The inefficiency problem

No other nation prices its gasoline so cheaply. At the very least, our gasoline should have a tax associated with it that is sufficient to rebuild our transportation infrastructure—just as other advanced nations do. Better still, the cost of our gasoline should reflect the cost of defending foreign oil fields, reversing the greenhouse effect, and solving the air quality issues. Best of all, the cost of our gasoline should include substantial funding for research on solar energy generation (and other renewable sources) and electric vehicle technologies—the two most environmentally beneficial and technologically promising gifts we can give to future generations.

A Proactive Solution

People living in the United States have been extremely fortunate. We have been blessed with clean air, abundant natural resources, a stable government, inexpensive energy costs, and a standard of living second to that of no other country on the planet. But nothing guarantees that our future generations will enjoy the same birthright. In fact, if we fold our hands behind our backs and walk away from today's environmental problems, we are guaranteeing that our children and our children's children will not enjoy the same standard of living as we do. For the sake of our children, we cannot walk away; we must do something. We must attack the problem straight on, pull it up by its roots, and replace it by a solution.

Figure 2-7 suggests a possible approach. We need to look at the results wanted in the mid-twenty-first century and work backward—on both the supply and the demand side—to see what we must start doing today. Clearly, it's time for a sweeping change—but we all have to want it and work toward it for it to happen. No one has to be hurt by the change if they become part of the change. Automakers can make more efficient vehicles. Suppliers can provide new parts in place of the old. The petrochemical industry can alter its mix to supply less refined crude as oil and gas and more as feedstock material used in making vehicles, homes, roads, and millions of other useful items. Long before any of these happens, you can do your part by building your own electric vehicle.

Today!!

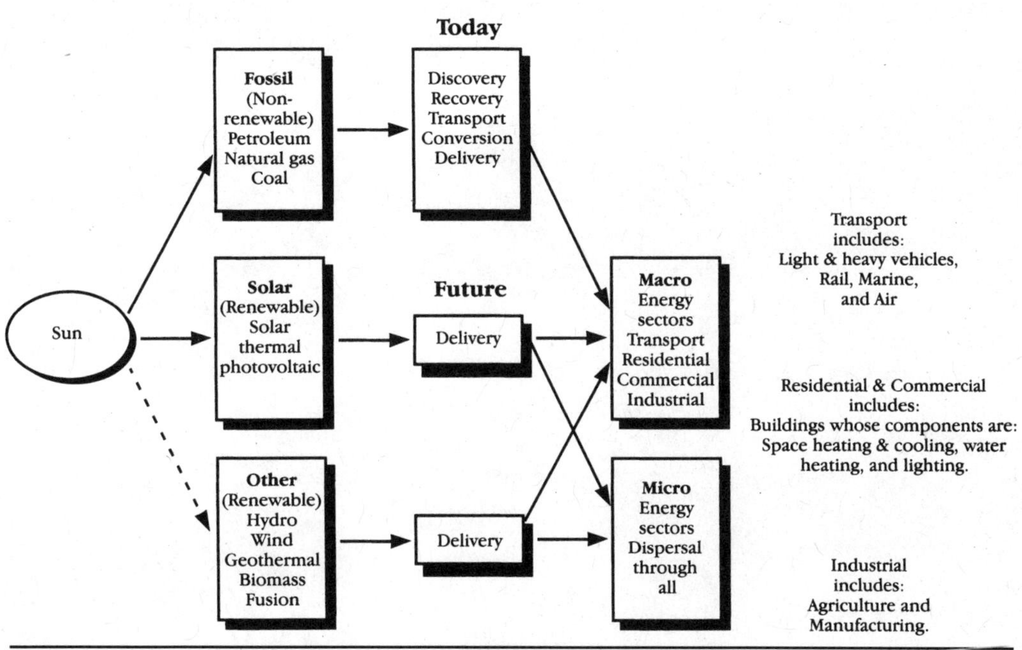

FIGURE 2-7 Model of balanced future energy usage made possible by working from a future desired goal back to today.

CHAPTER 3

History of Plug-In Hybrid Electric Vehicles

The term plug-in hybrid has come to mean a hybrid vehicle that can be charged from a standard electric wall socket.

Early Hybrid and Electric Vehicles

The production of hybrid electric vehicles began in 1899 with people like Dr. Ferdinand Lohner-Porsche.

Lohner-Porsche PHEV

At the age of 23, Lohner-Porsche built his first car, the Lohner Electric Chaise, in Germany. It was the world's first front-wheel-drive car. Porsche's second car was a hybrid, using an internal combustion engine to spin a generator that provided power to electric motors located in the wheel hubs. On battery alone, the car could travel nearly 40 miles.

The Introduction of Plug-In Hybrids

In July 1969, *Popular Science* included an article on the General Motors XP-883 plug-in hybrid. This concept commuter vehicle housed six 12-volt lead-acid batteries in the trunk area and a transverse-mounted DC electric motor turning a front-wheel-drive trans-axle. The gasoline-powered engine was connected to the trans-axle by a worm gear. The car could be plugged into a standard 110-volt AC outlet for recharging.[1]

The September 1975 issue of *Popular Science* featured a cover story on an experimental "turbo-electric" hybrid that "plugs in overnight for thrifty driving around town." Built by electronics engineer Harry Grepke, the vehicle used eight 12-volt truck batteries and a turbine genset. Grepke claimed that the car had an all-electric range of 50 miles.[2]

Electric and Hybrid Vehicle Research, Development, and Demonstration Act of 1976

The U.S. Congress enacted Public Law 94-413, the Electric and Hybrid Vehicle Research, Development, and Demonstration Act of 1976. Among the law's objectives was to work with industry to improve batteries, motors, controllers, and other hybrid electric components.

Partnership for a New Generation of Vehicles

During the 1990s, automakers embraced other vehicle technologies to various degrees. The Clinton administration announced a government initiative called the Partnership for a New Generation of Vehicles (PNGV) program. This program stimulated the development of gasoline hybrid vehicles by domestic automakers. In the program, the government worked with the American auto industry to develop a clean car that could operate at up to 80 miles per gallon. Several years and a billion dollars later, three prototypes for that 80-mpg car emerged. Every one of them was a hybrid.

While Detroit never actually deployed hybrid cars during this phase, the competitive spirit compelled Japanese automakers to do so. This led to popular vehicles like the Honda Civic hybrid and the Toyota Prius, with most major automakers eventually offering at least one hybrid model.

Among domestic automakers, hydrogen became the alternative fuel of choice for new concept cars, which were accompanied by promises to mass-market these vehicles by 2010. As we near the end of this decade, approximately 175 hydrogen fuel cell vehicles have been deployed in test fleets, but none have appeared in showrooms.

Who Really Killed the Electric Car?

Regulation in California

Legislation provided both the carrot and the stick to jump-start the development of electric vehicles (EVs). California started it all by mandating that beginning in 1998, 2 percent of each automaker's new-car fleet had to be made up of zero-emission vehicles (ZEVs), with the percentage rising to 10 percent by the year 2003. This would have meant 40,000 electric vehicles in California by 1998, and more than 500,000 by 2003.

California was quickly joined in its action by nearly all the northeastern states (ultimately, states representing more than half the market for vehicles in the United States had California-style mandates in place)—quite a stick! In addition, for Corporate Average Fuel Economy Standard (CAFE) purposes, every electric vehicle sold counted as a 200- to 400-mpg car under the 1988 Alternative Fuels Act.

But legislation also provided the carrot. California provided various financial incentives of up to $9,000 toward the purchase of an electric vehicle, as well as nonfinancial incentives, such as access to high-occupancy vehicle (HOV) lanes with only one person in the car. The National Energy Policy Act of 1992 allowed a 10 percent federal tax credit up to $4,000 on the purchase price of an EV. Other countries followed suit. Japan's MITI set a target of 200,000 domestic EVs in use by 2000, and both France and Holland enacted similar tax incentives to encourage electric vehicle purchase.

The California program was designed by the California Air Resources Board (CARB) to reduce air pollution, not specifically to promote electric vehicles. The regulation initially required simple "zero-emission vehicles" and didn't specify a required technology. At the time, electric vehicles and hydrogen fuel cell vehicles were the two known types of vehicles that would have complied; however, because fuel cells were (and are) fraught with technological and economic challenges, electric vehicles emerged as the technology of choice to meet the law.

The ZEV mandate in California created a backlash against electric vehicles by automakers, who didn't want to be required by legislation to build anything, let alone something other than their core internal combustion products. After the automakers banded together with the federal government and started legal proceedings against the state of California, the California Air Resources Board gutted the ZEV mandate (now known as the ZEV program), effectively releasing automakers from having to build electric vehicles at all. Under pressure from various manufacturers and the federal government, CARB replaced the zero-emission requirement with a combined requirement of a very small number of ZEVs to promote research and development and a much larger number of partial zero-emission vehicles (PZEVs), an administrative designation for super ultra low–emission vehicles (SULEVs), which emit about 10 percent of the pollution of ordinary low-emission vehicles and are also certified for zero evaporative emissions. While this was effective in reaching the air pollution goals projected for the zero-emission requirement, the market effect was to permit the major manufacturers to quickly terminate their public EV programs.

The years 2001 to 2005 initially marked a brief resurgence of internal combustion vehicles, particularly larger cars, trucks, and sport utility vehicles (SUVs). However, as public awareness of the issues surrounding petroleum dependence—climate change, political instability, and public health issues resulting from poor air quality, to name a few—has increased, the tide seems to be turning back toward plug-in vehicles. This has been stimulated both on a mass level by pop-culture movies such as *An Inconvenient Truth* and *Who Killed the Electric Car?* (discussed later), the number 1– and number 3–rated documentaries of 2006, respectively, and on a very personal level by rising gasoline prices. Increasingly, plug-in vehicles, which were once seen as a crunchy, environmental choice, are gathering bipartisan support, as those concerned with energy security are beginning to embrace the alternative of using cheap, clean, domestic electricity to power vehicles instead of expensive, foreign, comparatively dirty petroleum. With this broad coalition of support and declining auto sales, automakers have had little choice but to get onboard with newer alternatives to internal combustion vehicles.

New technology has also stimulated enthusiasm; in addition to electric vehicles, automakers have started working on low-speed electric vehicles, hybrid electric vehicles, and plug-in hybrid electric vehicles (PHEVs) (which combine a certain number of electric miles with the "safety net" of a hybrid propulsion system), and people have begun building their own electric cars. Depending on the vehicle's configuration, drivers might drive exclusively in electric mode for their weekly commutes, using gasoline only when driving long distances. This "best of both worlds" concept has renewed enthusiasm for electric vehicles as well, and both types of plug-in vehicles are benefiting from newer lithium ion battery technology, which stores more energy than previous lead-acid and nickel–metal hydride types, providing longer range.

September 11 and Our New Understanding of Electric Cars

Understandably, the attacks on the World Trade Center and the Pentagon on September 11, 2001, clearly show us that our reliance on imported oils is damaging our national and financial security. With oil reaching over $140 a barrel, the resurgence of electric cars, hybrid electric cars, and plug-in hybrid cars provides further evidence that we need to change and that there are immediate ways to change our ways.

A small selection of all-electric cars from the big automakers—including Honda's EV Plus, GM's EV1 and S-10 electric pickup, a Ford Ranger pickup, and Toyota's RAV4 EV—were introduced in California. Despite the enthusiasm of early adopters, the electrics failed to reach beyond a few hundred drivers for each model. Within a few years, the all-electric programs were dropped.[3]

Since the end of the electric car programs, the market has developed an expansive appetite for hybrid electric cars and cleaner gasoline cars. While many of the earlier electric cars, such as GM's EV1 (see Figures 3-1 and 3-2) and EV2, Chrysler's EPIC Minivan, and Ford's Ranger, as well as Honda's EV Plus, Nissan's Hypermini, and Toyota's RAV4, were recalled and destroyed, roughly 1,000 of these vehicles remain in private hands, as a result of both public pressure and campaigns waged by grassroots organizations such as "dontcrush.com" (now known as Plug In America), Rainforest Action Network, and Greenpeace.

Contradicting automakers' claims of anemic demand for EVs, these vehicles now often sell on the secondary market for more than they did when they were new. The whole episode became such a debacle that it spawned a feature-length documentary, directed by former EV1 driver and activist Chris Paine, entitled *Who Killed the Electric Car?*, which premiered at the 2006 Sundance Film Festival and was released in theaters by Sony Pictures Classics.

Hybrid Electric Cars Come to the Market

Toyota Prius

When Toyota released the Toyota Prius, it was the first hybrid four-door sedan available in the United States (see Figure 3-3).

Figure 3-1 The GM EV1 electric vehicles being crushed, as seen in *Who Killed The Electric Car?* Source: http://ev1-club.power.net/archive/031219/jpg/after2.htm.

Figure 3-2 The GM EV1 next to a Detroit electric car from 1915. Source: Russ Lemons.

FIGURE 3-3 The Toyota Prius. Source: Wikipedia.

The Toyota Prius II won the 2004 Car of the Year Awards from *Motor Trend* magazine and the North American Auto Show. Toyota was surprised by the demand and pumped up its production for the U.S. market from 36,000 to 47,000. Interested buyers waited up to six months to purchase the 2004 Prius. Toyota Motor Sales U.S.A. President Jim Press called it "the hottest car we've ever had."[4]

The Toyota Prius was introduced to the Japanese market two years before its original launch date and prior to the Kyoto conference on global warming held in December of that year. First-year sales were nearly 18,000.

Honda Insight

Honda released the two-door Insight, the first hybrid car to hit the mass market in the United States. The Insight won numerous awards and received EPA mileage ratings of 61 mpg city and 70 mpg highway.

Honda Civic Hybrid

Honda introduced the Honda Civic Hybrid, its second commercially available hybrid gasoline-electric car. The appearance and drivability of the Civic Hybrid were (and still are) identical to those of the conventional Civic.

Ford Escape Hybrid

In September 2004, Ford released the Escape Hybrid, the first American hybrid and the first SUV hybrid.[5]

Though carmakers now speak positively of PHEVs, it could be years before you'll find them at your local dealership.[6] If you want one now, you'll spend far more than they'll cost when they are mass-produced. Today, you'll buy one to be among the first to drive the world's cleanest extended-range vehicle; in the future, you'll do it because it's an obvious and affordable win!

PHEVs Come to Market Thanks to Andy Frank (the Father of PHEVs)

In 1971, Dr. Andy Frank (as seen in Figure 3-4), the inventor of the modern PHEV, began working on hybrids and PHEVs. He is professor of mechanical and aeronautical engineering at the University of California at Davis.

Beginning around 1990, Professor Frank began using student teams to build operational prototype plug-in hybrid electric vehicles. The UC Davis PHEVs won several DOE/USCAR "Future Car" and "Future Truck" national competitions.

Inspired by his work as an EV1 propulsion system engineer, Jeff Ronning began developing concepts for plug-in hybrids in the mid-1990s at what was then the Delco Remy division of GM. EV1 prototypes sometimes had attached "range-extender" trailers, developed by Alan Cocconi of AC Propulsion. These trailers

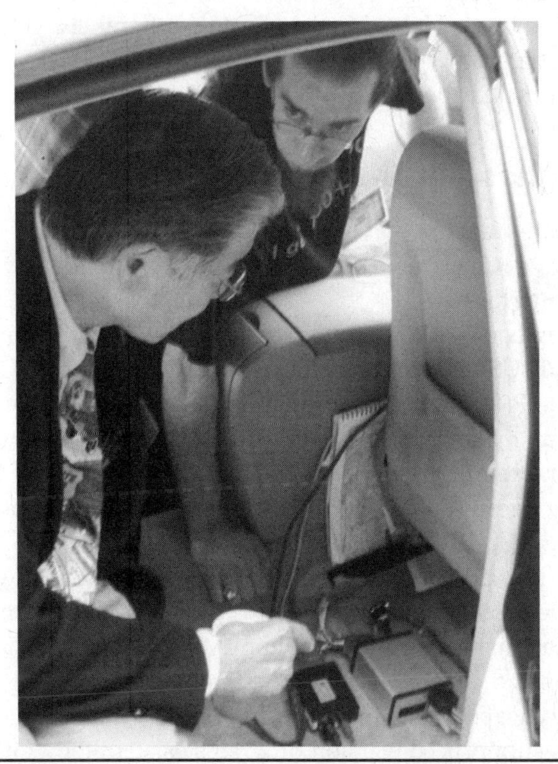

FIGURE 3-4 Dr. Andy Frank and Ryan Fulcher inside a converted Toyota Prius. Source: CalCars.org.

were simply rolling generating sets that could supply power for long trips. It was only natural to conceive of an EV1 with a small turbo-alternator on board (1995 internal publication). External publications (SAE 971629 and 1999-01-2946) followed expounding the merits of using electric energy for most local travel and proving it with the data from the U.S. DOT.[7] "Unlimited EV," "battery dominant hybrid," and "energy hybrid" were the names given to the architecture because the term plug-in hybrid was coined by Dr. Andy Frank.

By the end of the year, Dr. Frank's student teams had built and operated seven proof-of-concept and proof-of-demonstration prototype plug-in hybrid electric vehicles, including six-passenger sedans (Taurus and Sable), sport utility vehicles (Suburban and Explorer), a two-seater sports car (GM EV1), and two ground-up 80-mpg sports cars, and the CalCars PRIUS+ prototype and EDrive Systems conversions were demonstrated.[8]

Hybrid Center at the University of California–Davis

Under Professor Frank, this academic program has converted a number of internal combustion engine vehicles for demonstration purposes. Professor Frank has formed Efficient Drivetrains, Inc., to put his intellectual property to work with carmakers and suppliers.

At the UC Davis Hybrid Center, Professor Frank leads the teams to design and build working prototypes that work. In 2006, they did a PHEV conversion of a GM Equinox for the Challenge X competition.[9]

The California Cars Initiative (CalCars)

In 2002, entrepreneurs, environmentalists, and engineers created the California Cars Initiative (CalCars). CalCars is a nonprofit PHEV advocacy and technology development group.

In September, CalCars converted a 2004 Toyota Prius into a prototype of what it calls the PRIUS+ (see Figure 3-5). With the addition of 130 kg (300 lb) of lead-acid batteries, the PRIUS+ achieved roughly double the gasoline mileage of a standard Prius and can make trips of up to 15 km (10 miles) using only electric power.[10]

Felix Kramer, the founder of CalCars, is the owner of the PRIUS+ in Figure 3-5, thereby becoming the world's first consumer owner of a plug-in hybrid (see Figure 3-6).

Kramer put 25,000 miles on the car in the year after it was converted in April 2006, and flew it to Washington to give rides to senators and representatives and put the car at their service, thus building what former CIA Director James Woolsey describes as the "coalition between the tree-huggers, the do-gooders, the sod-busters, the cheap hawks, and the evangelicals."[11]

CalCars supports conversion programs as a strategy for increasing awareness of and support for PHEVs and thereby motivating automakers to build production

FIGURE 3-5 The first PRIUS+ conversion, September 11–22, 2004. The team started with an initial lead-acid battery pack to prove that the system worked. Pictured are (left to right) Ron Gremban, Felix Kramer, Marc Geller, Kevin Lyons, and Andrew Lawton. Others who helped included Tom Driscoll, Michael Geller, Richard Jesch, Les Montavon, Dan Putman, Michael Schwabe, Len Tramiel, Bob Westman, and Doug Widney. Courtesy of Andy Frank and CalCars (www.calcars.org).

FIGURE 3-6 Felix Kramer is the world's first consumer owner of a plug-in hybrid.

PHEVs. Though costs are now high, we believe that automakers could sell PHEVs for only $2,000 to $5,000 more than current hybrids.

CalCars Demonstration Conversions

The nonprofit CalCars does conversions to demonstrate new designs and provide a platform for different batteries; it doesn't sell conversions. CalCars has sponsored the EAA-PHEV project, and its open-source designs are being used by some of the private companies as well as by technically advanced individuals.[12]

CalCars is continuing to evaluate new batteries of different types and to give battery companies a platform to demonstrate their products. Automakers are watching these efforts, and these projects could result in the group's attracting a battery maker to provide batteries to the EAA-PHEV project members for a group price.

In January 2007, CalCars began a project to install Valence batteries in several cars. Earlier, in 2006, CalCars worked on a project with Electro Energy, Inc., to demonstrate its PRIUS+ approach using nickel–metal hydride batteries, in response to automakers' continuing comments that "the batteries aren't ready." This project demonstrated that the battery type found in all current hybrids can make a well-performing PHEV. We believe that both nickel–metal hydride and lithium ion batteries are ready.

Plug In America

Plug In America (also known as PIA) is a nonprofit educational organization that promotes and advocates the use of plug-in cars, trucks, and sport utility vehicles (SUVs) powered by domestic electricity, which it claims will help reduce dependence on oil and improve the global environment. It is currently a chapter of the Electric Auto Association. The Managing Director is Jeanne Trombly.

PIA advocates the development of plug-in hybrid electric vehicles, battery electric vehicles, and other vehicles that utilize electricity, either from the power grid or from electricity-generating devices such as solar cells, as a substantial source of motive energy.[13]

EPRI-DaimlerChrysler Plug-In Hybrid Development Program

Back when I worked for the New York Power Authority, I remember that my boss, Bart Chezar, told me that he was working on a report for the Electric Power Research Institute (EPRI) to support the commercialization of PHEVs. Other sponsoring entities were automakers, national labs, other utilities, and the University of California at Davis. This group even received support from the European Commission (EU) and other government agencies worldwide.

When the Whole Study Started in 2001

The U.S. Department of Energy (U.S. DOE) created the National Center of Hybrid Excellence at UC Davis, with Dr. Frank as director. Dr. Frank also obtained substantial GM funds to hybridize and plug in GM's EV1.

DaimlerChrysler, working with the Electric Power Research Institute (EPRI) and others, has produced a small number of PHEV Dodge Sprinter 15-passenger vans for fleet testing purposes. Additional DaimlerChrysler Mercedes-Benz Sprinter 15-passenger van PHEV prototypes were also completed. EPRI, along with a number of utilities (such as the New York Power Authority) and government agencies, worked with DaimlerChrysler to deliver four Sprinter PHEV vans to test fleets.[14]

As seen in Figure 3-7, the PHEV hybrid Sprinter program even received support from then-President Bush and the car companies.

In August, four companies—Raser Technologies (www.rasertech.com), Maxwell Technologies, Electrovaya, and Pacific Gas and Electric—formed the Plug-In Hybrid Consortium to help reduce the research and design gap between component suppliers and OEMs and to accelerate the development of critical PHEV components. Since then, nine other component companies and three more utility companies, as well as CalCars and Plug In America, have joined the consortium.[15]

CARB Support for PHEVs

On March 27, 2008, CARB modified its regulations, requiring car companies to produce 58,000 plug-in hybrids for sale to Californians between 2012 and 2014.[16]

Economic Stimulus for PHEVs

On October 3, the United States enacted the Energy Improvement and Extension Act of 2008 as part of the Emergency Economic Stabilization Act of 2008.

Figure 3-7 President Bush with a PHEV Dodge Sprinter Van as converted in the EPRI program.

President Obama has signed the stimulus bill, which has authorized more than $2 billion for plug-in technology. This will unquestionably put more numbers and kinds of plug-in electric vehicles on the road. It will help create jobs and spur spending by giving consumers incentives to purchase the cleanest-running vehicles made today and those that are just around the corner.[17]

The new tax credits for plug-in vehicles will range from $2,500 to $7,500, with factors such as battery capacity determining how much owners will receive. The total cost of the program over the next 10 years is estimated at $2.8 billion—a significant sum of money, but a drop in the bucket next to the $700 billion bill it's a part of or the money received so far by Chrysler and GM.

Credits are to be capped based on vehicle size. Consumer vehicles such as the Volt will get a maximum of $7,500. Lighter commercial trucks and vans will get up to $10,000. The heavy Azure vans could get up to $12,500.

To get $15,000, vehicles will have to weigh more than 26,000 pounds, well into the gargantuan commercial vehicle range. They will also have to have a nearly 1,000-pound battery pack—large enough to power the average home in northeastern Ohio for a day.[18]

To meet the tax incentive's standards, a plug-in vehicle must have a battery with a minimum capacity of 4 kWh, although an additional $200 tax credit is added for every kilowatt-hour thereafter, which is how the Volt gets to the maximum $7,500 limit with its 16-kWh battery.[19]

Conversion Companies

There are a number of companies that provide conversions to plug-in hybrids.[20]

Hybrids Plus PHEV

Hybrids Plus of Boulder, Colorado, began offering plug-in hybrid conversions of the 2004, and later the Toyota Prius, using A123 lithium ion batteries for either a 15- or a 30-mile all-electric range. Its first contract was with the Colorado Office of Energy Management and Conservation, which ordered one plug-in Prius conversion. The vehicle was handed over to Colorado OEMC on March 6, 2007. Current pricing starts at $24,000.

Hybrids Plus also won a contract to convert a Ford Escape Hybrid to a plug-in hybrid for one of the public authorities that I once worked for, the New York State Energy Research and Development Authority.[21]

A123Systems and Hymotion

A123Systems

Founded in Hopkinton, Massachusetts, during 2001, A123Systems proprietary Nanophosphate technology builds on new nanoscale materials initially developed at the Massachusetts Institute of Technology (MIT). A123Systems is now one of the world's leading suppliers of high-power lithium ion batteries, using its patented Nanophosphate technology to deliver a new combination of power, safety, and life.[22]

A123Systems CEO David Vieau announced in 2007 to the U.S. Senate Committee on Finance Subcommittee on Energy, Natural Resources, and Infrastructure that the company planned to market battery packs for third-party conversion of hybrids to plug-in hybrids in 2008.[23]

Then came Hymotion.

Hymotion

Hymotion, originally started in Concord, Ontario, introduced plug-in hybrid upgrade kits in February 2006. Designed for the Toyota Prius and the Ford Escape and Mariner hybrids, these kits were offered first to fleet buyers and are projected to be available to the general public in 2007.

The Hymotion PHEV module requires minimal modification to the stock vehicle. All necessary components and safety features are integrated and contained within the module, including batteries, power electronics, crash sensors, charger, battery management system, safety sensors, and manual-electric interlock. The system does not require removal of the OEM battery pack and can be installed in less than two hours, according to the company.

Since its acquisition by battery maker A123Systems, Hymotion crash-tested Prius conversions can now be ordered for $9,950 plus $400 delivery, with a $1,000 deposit required, from distributors in several cities. Go to Hymotion.com. Green Gears, Inc., is Hymotion's San Francisco–based distributor/installer.[24]

The PHEVs Available or Soon to Come onto the Market

We know that there are a bunch of these, but we need to look at all the various PHEVs that are on the market or soon to be available. More importantly, you can convert a car to a PHEV for $8,000 or less, using off-the-shelf parts.

This section will look at PHEVs from the various automakers, conversion companies, and soon-to-be companies that will be coming on the market.

Toyota's Plug-In Hybrids Jumping In from Japan (First)

In July 2008, Toyota Motor Corporation (Toyota), which already has the best hybrid in town, announced that it would have a plug-in hybrid available. Toyota is even proposing a 90-plus-mpg PHEV.

Toyota expects to have a plug-in hybrid vehicle on the market by 2010. The company had said earlier that it would market lithium ion batteries, but now it has somewhere to put them. Felix Kramer, founder of CalCars, called Toyota's announcement "stunning and very welcome."

Early versions of Toyota's plug-in car, based on the Prius, are being tested at two campuses of the University of California. These preproduction cars can operate in all-electric mode for only 7 miles, but the 2010 vehicle will probably bump that up to 20 to 30 miles. The current plug-in switches back to the gas motor at 62 miles per hour.

According to Toyota President Katsuaki Watanabe in Detroit, "By 2010 we will accelerate our global plug-in hybrid R&D program. As part of this plan, we will deliver a significant fleet [of plug-ins] powered by lithium-ion batteries to a wide variety of global commercial customers, with many coming to the U.S."[25]

GM and the Chevy Volt

How Does It Work?

One of the engineers from GM explained to me how the Volt works. It is basically a 40-mph/40-mile electric car. After that, it changes into a gas car—basically a series hybrid.

While a series hybrid is good, the Toyota Prius gets 100 miles per gallon because after the 20 miles of pure EV mode, it changes into a hybrid electric car. I personally like the hybrid car after the pure EV, but who am I to judge?

Economics

The reason the automakers need to make this technology work is an economic one: it would cost less than $1.00 a day to charge up! Yup, we are talking about no charging stations. Using a Society of Automotive Engineers 1772 standard for chargers makes the electric car more versatile and safe, since you are using a 110/120-V outlet. That allows the driver to charge up while at work or at the train station. What can that do? Turn the car into a 80-mile-range EV.

City Driving—All Electric

Yes, all city driving would then be virtually all electric. That is an emission-reduction dream and the ability to make every car in the GM family a plug-in hybrid. Looking at the idea of a plug-in hybrid electric car in terms of fuel efficiency standards shows how some car companies would choose to go all electric.

The range-extended EV provides the consumer with a great start for an electric vehicle. While I personally might like a plug-in hybrid that then goes to hybrid, to get the market ready for the technology, an EV range/dual-fuel car can work.

GM should add the Volt technology to every car. The GM Volt (range-extending EV) comes in at 50 miles per gallon. Toyota's is basically at 100 miles per gallon. If you charged up the Toyota Prius PHEV, you would get 200 miles per gallon and a 40-mile EV range, the same as that of the Volt.

Going Backward before Going Forward

One reporter at the 2008 General Motors press event for the Chevy Volt asked a GM manager how, if there is no hybrid electric vehicle after 40 EV miles, will that make

Figure 3-8 The display for the Series Hybrid System of the GM Volt.

a car worth over $40,000. Figure 3-8 shows the display for the Series Hybrid System of the GM Volt. The display clearly shows how many miles you would drive all-electric and all-gas.

Like the reporter for *Forbes* magazine, I kind of wonder that, too, since a PHEV should allow for an all-electric range, then kick into a hybrid to get more energy savings for your buck. That is what really makes the car more cost-effective, more efficient, and more environmentally benign. In addition, you would think that a dual-fuel EV/gas car would be more cost-effective than the Prius. Guess not.

The Electric Part of the Volt Is What People Want!

This brings me back to the electric car. You see, all we are doing is going back to what we started from, which is the essence of the automobile (since the first cars were electric).

While hybrids have recently been criticized for not being as efficient or as cost-effective as they could be, plug-in hybrids (and, more importantly, pure EVs) could be much more competitive if all vehicles in the fleet each year were partially battery, whether hybrid or dual fuel. Why? Purchasing more batteries at a single time reduces the overall cost of the batteries for the car companies. This will reduce the overall cost of the car, making plug-in hybrids more competitive.

Figure 3-9 Rear view of the GM Volt.

What Could They Do?

Here are some other important aspects of pure electric cars (or the purely electric part of a PHEV).

The three biggest factors (always regarding electric cars) that could possibly help the Big Three (I hope) get out of this mess are the following:

1. Electric cars are zero-emission vehicles. Also, power plants will reduce their emissions over time, making electric cars cleaner once the stationary source that powers the vehicles gets cleaner.
2. More important, if renewable energy sources such as wind, solar, geothermal, and tidal energy, rather than coal or other unsustainable energy sources, become a larger part of the energy portfolio, the vehicles will be fully zero emission and oil free!
3. Electric cars cost pennies to charge. In addition, there is vehicle-to-grid (V2G), which is when the recharging of the electric car's batteries (energy storage) reverse-meters your energy costs. Thus, when you recharge, the utilities take energy from your car at 15-minute intervals, and give you a credit on your energy bill.

FIGURE 3-10 Side view of the GM Volt with the plug . . . the best part.

So, What Is There to Do?

Evolve. Electric cars, plug-in hybrids, and hybrids are the solutions for revolutionizing the auto industry. Therefore, the automakers should stop delaying production of these cars. In fact, they should accelerate production of the Volt.[26]

Fisker Karma Plug-In Hybrid Sport Sedan

Another kind of plug-in hybrid shown in Detroit is the luxury four-door Fisker Karma sport sedan (see Figure 3-11), which certainly makes some strong claims, including zero to 60 in less than six seconds and 50 miles of all-electric range. It is said that this car will be on the market in 2009 for $80,000, with annual sales of 15,000 projected. We'll believe it when we see it, because the road to independent car production is so rough. The media apparently thought that the California-based Tesla (which had "cool" high-tech connections, millionaire investors, and a gorgeous sports car as part of the package) was different. But now its electric sports car dreams seem to be unraveling amid layoffs and missed deadlines.

AFS Trinity's "Extreme Hybrid"

Then there is the "Extreme Hybrid" from New York–based AFS Trinity (see Figure 3-12). "Amazing," said the reporter. "It gets 150 mpg, with 40 miles of all-electric range!" But the car is a plug-in conversion of the existing Saturn Vue hybrid (in other words, a homemade version of the same car GM says it will actually build).

Figure 3-11 The Fisker Karma.

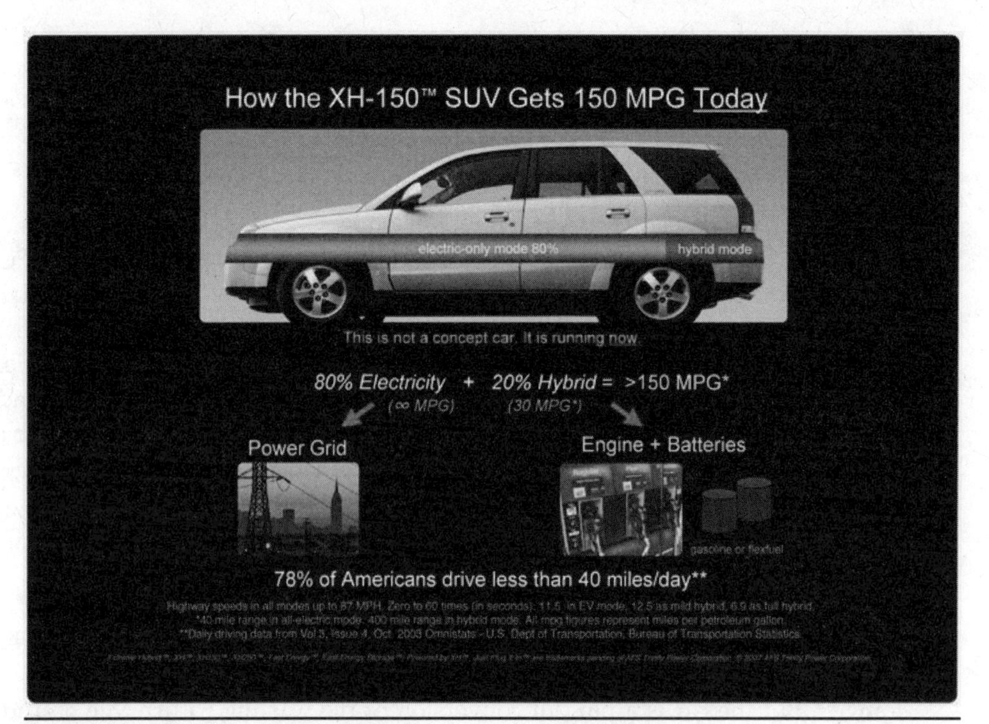

Figure 3-12 AFS Trinity "Extreme Hybrid" for a plug-in Saturn Vue.

Other aftermarket plug-ins are on the market, but I would wait for the carmakers in 2010. AFS says that it wants to license its technology to the industry, but that road is paved with tragedy.[27]

Ford Motor Company and the Escape PHEV

Ford has developed a small fleet of plug-in hybrids, but it has no plans to offer them to the public.[28]

Phoenix Motorcars

In October 2008, Phoenix Motorcars and UQM Technologies announced a project to develop a plug-in series hybrid version of the sport utility truck that Phoenix currently sells as an all-electric vehicle. The vehicle is projected to use a small gasoline-fueled internal combustion engine as a range extender and lithium titanate batteries from Altair Nanotechnologies.[29]

Conclusion: PHEVs Are Here Today!

We use this section to track the existing plug-in hybrids, and to prove definitively that PHEVs can rely exclusively on existing technology—no new advances are required. PHEV conversions are emerging at a frenzied rate (see the frequency of CalCars–News postings), to the point where it's no longer feasible to track every instance of a PHEV. In short (and according to CalCars)[30]:

- In 2003–2004, the U.S. Marine Corps demonstrated a diesel-electric PHEV-20 HUMVEE. This advanced Shadow RST-V (Reconnaissance, Surveillance and Targeting Vehicle) PHEV, built by General Dynamics, uses lightweight lithium ion batteries and motors in four wheel hubs.
- Several companies are building plug-in hybrid school buses. And in Long Island, New York, a company has converted a city bus to a plug-in hybrid with 40 miles of all-electric range. Many more heavy-duty vehicle conversions (including three recycling dump trucks that will run in "silent" mode for pickups) are in progress.

Game Changers

While electric cars have a long-standing place in our automotive history, plug-in hybrids have a short history that is rich and promising. You see as Scherry Boschert once wrote in *Plug-In Hybrids: The Cars That Will Recharge America*,[31] "Many, many people are integral to the story of plug-in hybrids."

These include:

- Chelsea Sexton, automotive insider. Working for General Motors, Sexton fought attempts to destroy the electric EV1 car. Her experience illustrates how car companies are resisting plug-in cars, and why they'll make them anyway.

- Felix Kramer and the tech squad. Geek power, in all the best meanings of the phrase, put plug-in hybrids on the public map. Hackers with expertise in computers and cars turned a Toyota Prius into a 100-miles-per-gallon (160-km-per-gallon) plug-in-hybrid and brought it to the attention of the world. What they did, the car companies will do even better, and on a much larger scale.
- Marc Geller, grassroots activist. Inspired by the successful protests of Act Up, Queer Nation, and antiwar activists, Geller helped organize street demonstrations that shamed some car companies into ceasing destruction of electric vehicles. The actions put the lie to automakers' claims that nobody wants plug-in cars, and helped pave the way for plug-in hybrids.
- R. James Woolsey, former CIA director and national security hawk. Seeing the end of cheap oil supplies looming, Woolsey advocates for plug-in hybrids to wean us from petroleum and to divert former "petrodollars" away from Islamic radicals.

Conservatives in high office during the Bush administration and now the Obama administration are being influenced by his arguments.

Now, while Sherry and I do not necessarily agree with everyone's opinions represented here, it is a game changer to have all these segments of the political sphere and others support plug-in hybrids together. That makes the issue not just the "tree huggers" or the environmentalists' sole issue to coddle. The hawks like it for its reduction of our reliance on imported oil. Also, I believe that since the idea of a plug-in hybrid is being richly pursued internationally, the United States (as always) is moving the ball forward on the technology.

PHEV Conversions

PHEV conversions are limited to certain vehicle types. So far, most vehicles that are converted start as hybrids. The easiest conversions are those for the 2004–2008 Prius (not the 2001–2003 Prius) and the Ford Escape/Mercury Mariner hybrid. Because of Honda's different architecture (the engine runs whenever the motor runs), aftermarket conversion of Hondas is unlikely. (Both Toyota and Ford are now producing a few prototype conversions of their own Priuses and Escapes for research purposes.)

CHAPTER 4

What Is the Best Plug-In Hybrid Electric Vehicle for You?

The benefit of building/converting your own electric vehicle is that you get this capability at the best possible price with the greatest flexibility.
Bob Brant

Plug-In Hybrid Electric Vehicle Purchase Decisions

When you go out to buy a plug-in hybrid electric vehicle today, you have three choices:

- Buying a ready-to-run PHEV from a major automaker (you can't get that yet)
- Buying from a PHEV conversion shop
- Buying a used PHEV conversion from an individual

Your two most basic considerations are

- How much money you can spend
- How much time you have

Another element of your purchase trade-off is where (and from whom) you can obtain an electric vehicle. This section will present your options and highlight why conversion is your best choice at this time.

Conversions Save You Money and Time

There are several reasons why conversions are good.

A conversion

- Costs less money than either buying the car ready-to-run or building it from scratch.
- Takes less time than building the car from scratch and only a little more time than buying it ready-to-run.

You could typically spend from $12,500 to $15,000 for a new PHEV at an independent dealership. In only a month or two, you can be driving around in your own PHEV conversion for under $9,000. When you build a PHEV from your own car, your electrical parts cost the same and there are no labor costs. It is harder, and it is not for everyone, but if you can convert your car, do it!

Buying Ready-to-Run

Ready-to-run plug-in hybrid electric vehicles are not available today from any major automaker. Not even ten years ago, the major automakers had PHEVs available for lease and sale. Today, most automakers are making hybrids, and a few have pure PHEVs in various stages of development and testing. This includes GM's Volt, which I saw at an event for *Electrifying Times* in 2008. Since the Volt is not yet available, when I saw a picture for a Saturn Vue PHEV (Figure 4-1), I thought that

Figure 4-1 Saturn Vue PHEV.

might be a better idea. A picture of a Saab ethanol PHEV from General Motors was also shown on *EV World*. It is not yet available.

The bottom line is, there are a few *new* automakers that are *developing* new PHEVs.

Buying a Ready-to-Run Converted Toyota Prius from an Independent Manufacturer

Ready-to-run Prius plug-in hybrid electric vehicles are more readily available from independent manufacturers and their dealers today. Here are some examples.

Hymotion is one great example. With the backing of A123 Systems (we'll get more about this company in Chapter 8, "Batteries"), Hymotion developed a great PHEV conversion system and established a dealership network across the country; its conversion system is meant for the second-generation Toyota Prius.

As we know, PHEVs have the potential to be far lower in cost than internal combustion–powered vehicles in the future, when economies of scale from increased production kick in, because there are far fewer components that go into an electric vehicle than into an internal combustion–powered vehicle. In addition, as the price of gas increases over time, the cost incentives will increase as well.

Many PHEV conversions use the 2004 Toyota Prius for model years 2004 and earlier. Some of the systems have involved replacement of the vehicle's original nickel–metal hydride battery pack and its electronic control unit. Others, such as Hymotion's, the CalCars PRIUS+, and the PiPrius, piggyback an additional battery pack onto the original battery pack; these are also referred to as battery range extender modules (BREMs).

Within the EV community, this is called the "hybrid battery pack configuration." Also, most of the original conversions started by CalCars used lead-acid battery conversions. Those vehicles got a 10-mile EV range and a 20-mile blended range.

Conversions of Production Hybrids

Converting Existing Hybrid Electric Vehicles

Conversion of an existing hybrid electric vehicle is the best alternative because it costs less than either buying a PHEV ready-made or building one from scratch, takes only a little more time than buying ready-made, and is technically within everyone's reach (certainly with the help of a local mechanic, and absolutely with the help of a PHEV conversion shop).

Conversion is also easiest from the labor standpoint. You buy the existing hybrid you like (certain chassis types are easier and better to convert than others), put an electric motor in your chassis, and save a bundle. It's really quite simple; Chapter 10 covers the steps in detail.

Figure 4-2 shows a smart PHEV conversion using the AFS Trinity system.[1]

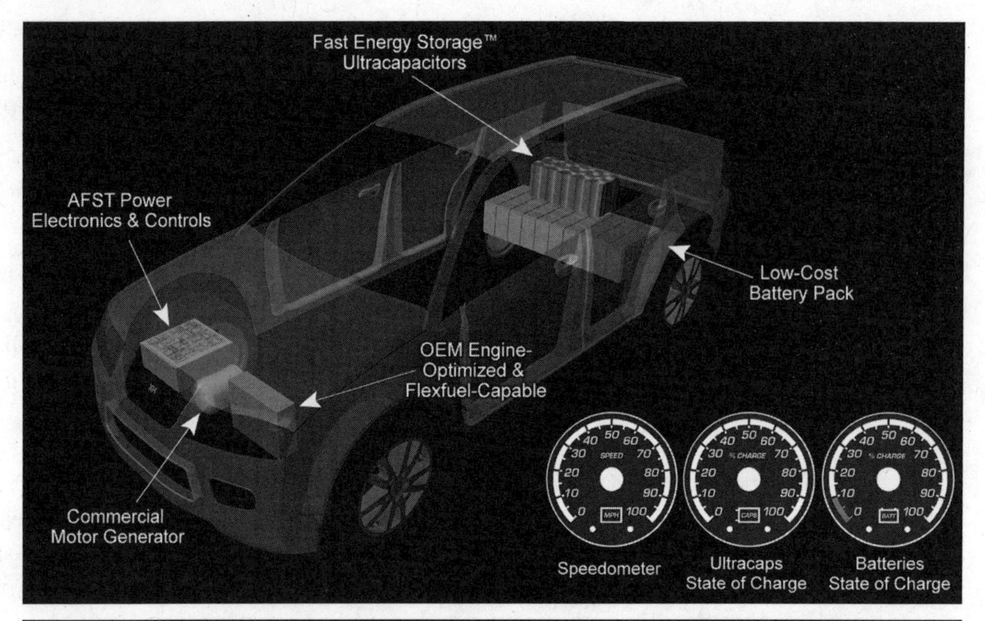

FIGURE 4-2 A 3D cutaway image of AFS Trinity's XH-150 plug-in hybrid electric car, including the speedometer, ultracapacitor, and battery state-of-charge gauges. Source: AFS Trinity; www.afstrinity.com/press-images.htm.

Figures 4-3 and 4-4 show 3D cutaway images of AFS Trinity's XH-150 plug-in hybrid electric car during initial charging from a household power plug. This "extreme hybrid" travels 40 miles in electric-only mode before the gas engine is used to go further. Since 78 percent of Americans drive less than 40 miles per day, this would enable many people who might own an XH-150 never to have to visit a gas station.[2]

Figure 4-5 shows a 3D cutaway image of AFS Trinity's XH-150 plug-in hybrid electric car during acceleration. By using ultracapacitors to buffer and protect its lithium ion batteries during acceleration, the XH-150 is able to achieve fast acceleration but avoid excessive resistive heating.

There are a number of conversion shops around the country. Conversion shops will convert the vehicle of your choice. In many respects, this is an advantage because you get the vehicle you want converted by a professional. On the Plug In America and CalCars websites, you will get a great list of them (as well as in the source directory in Chapter 12).

Some Conversion Examples

Figure 4-6 shows the fully electric Hi-Pa Drive F-150 concept vehicle, which uses the Ford F-150—the world's bestselling pickup truck—to demonstrate what

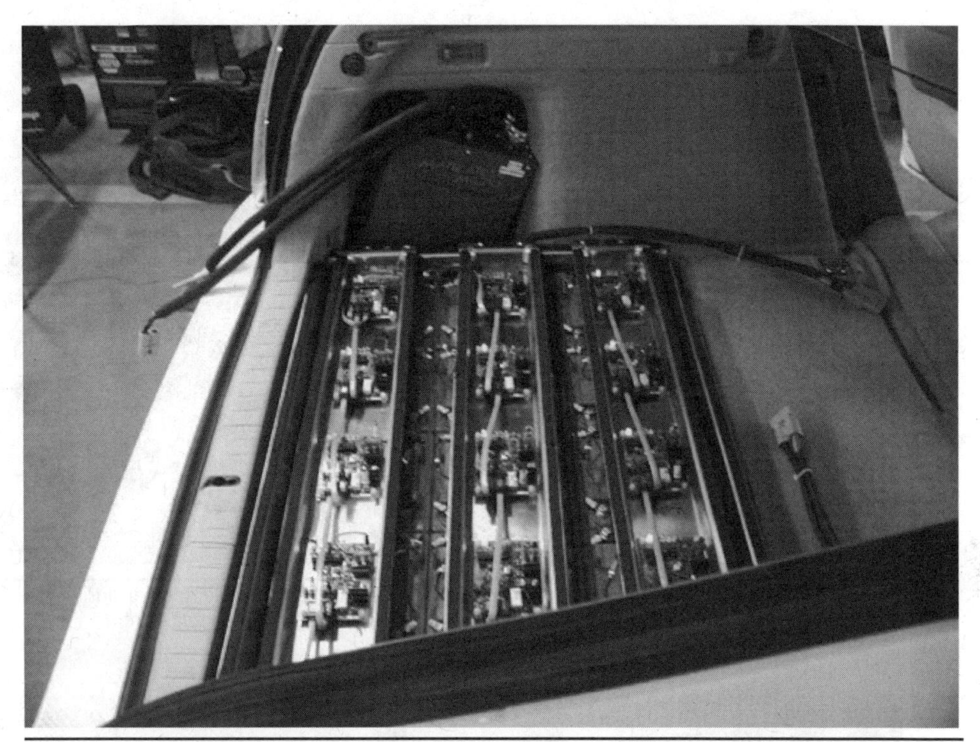

Figure 4-3 AFS Trinity XH-150 PHEV during charging. Source: AFS Trinity; www.afstrinity.com/press-images.htm.

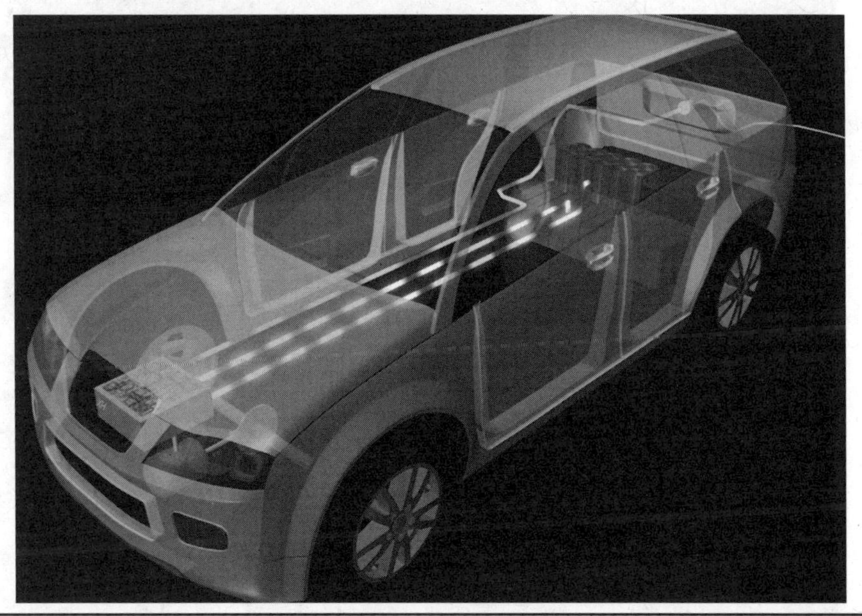

Figure 4-4 A 3D cutaway image of AFS Trinity's XH-150 plug-in hybrid electric car during charging from a household power plug (front). Source: AFS Trinity;www.afstrinity.com/press-images.htm.

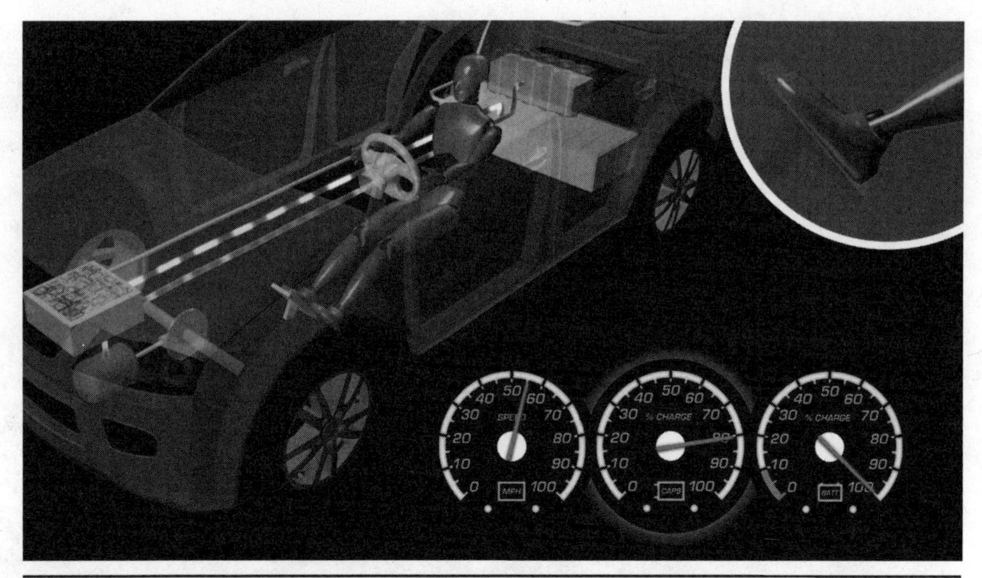

Figure 4-5 A 3D cutaway image of AFS Trinity's XH-150 plug-in hybrid electric car during acceleration. Source: AFS Trinity; www.afstrinity.com/press-images.htm.

Figure 4-6 PHEV 100 with the Ford F-150.

automobile manufacturers and after-market innovators are capable of doing today to profitably create vehicles that are powerful, clean, economical, and fun to drive. The concept vehicle was built in collaboration with Ford by DST industries and PML, the creators of the revolutionary Hi-Pa Drive system that powers the truck. The Hi-Pa F-150 truck is powered by four Hi-Pa Drive in-wheel motors that will be able to deliver more than 600 horsepower where it matters most, and recapture most of the energy used in braking.[3]

The Hi-Pa F-150 Wheel Motor shows the ability to place additional electric motors into the system (shown in Figure 4-7). The vehicle shown has a small ICE motor, an electric motor, a controller, battery packs, and also standard gauges, clips, and so on. Notice the neatly laid out components and how they tie into the engine.

Figure 4-8 shows a converted Volkswagen Rabbit PHEV. In 2003, AC Propulsion (the guys who developed the drive train for the EV1) converted a four-passenger Volkswagen Jetta to PHEV. This car can use either gas or natural gas to power the lead-acid batteries.

This car also used vehicle-to-grid (V2G) technology, whereby the utilities can use the car as solar panels do by reversing the direction of your electric meter.

According to tests conducted by the California Air Resources Board (CARB), the Electric Power Research Institute (EPRI), the South Coast Air Quality Management District, Volkswagen, and the National Renewable Energy Laboratory (NREL), this PHEV has a 30-mile ZEV range.[4]

FIGURE 4-7 Hi-Pa F-150 Wheel Motor. Source: PML FlightLink.

Figure 4-8 Volkswagen Rabbit PHEV.

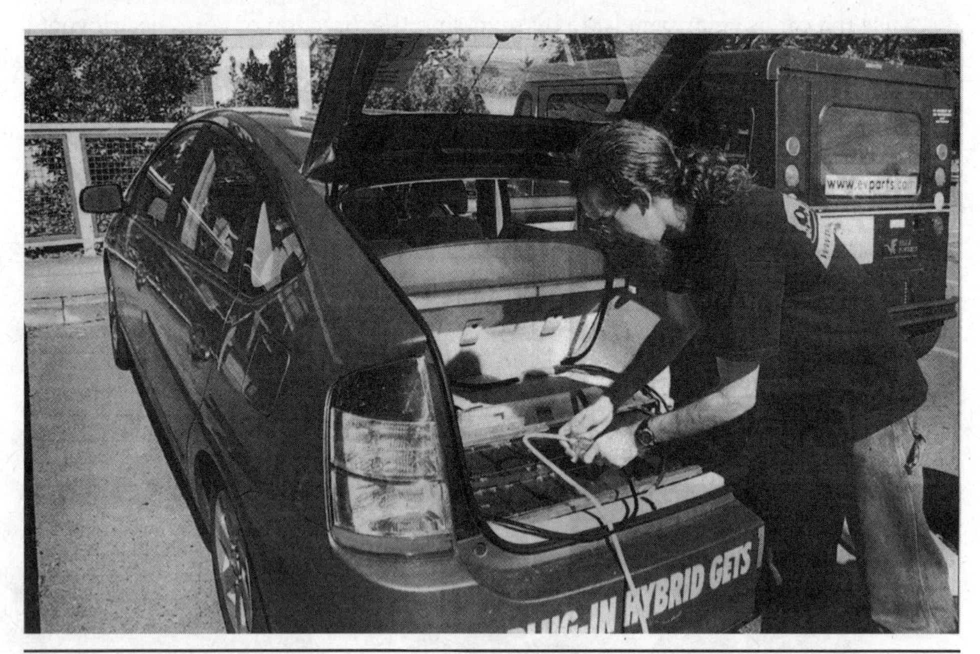

Figure 4-9 The Manzanita Micro Toyota Prius PHEV. Source: Manzanita Micro.

Table 4-1 covers the other dimensions of the buy, convert, or build trade-off: what vehicles or parts are available, whom you can get them from, where these sources are located, and when you will get/assemble them. There are not many places to buy new PHEVs today.

TABLE 4-1 Electric Vehicle Purchase Decisions Compared http://www.eaa-phev.org/wiki/Prius_PHEV#Kits_and_Conversions

Organization: Location Websites (Products)	Conv. service	Conv. kit	Status	# done so far	EV range [mi]	PHEV range [mi]	AC power	Charge time [hours]	Safety	Added weight [kg]	Spare Tire	Cost [US$]	Warr. [years]	Type	Pack energy [KWh]	DOD energy [KWh]	Bat type
PriusPlus:CA [10] CalCars (PbNiMH)	No	Yes	Dev Doc	5	10–12 20–25	20+ 40+	100 to 240 Vac	4+ 5	Flame Spill proof	130 100	Opt	$3-$9K +Labor	0	Hyb Con CV	4.8 6.5	2.4-3.8 +0.3 stock 5	PbA (Ni, Li)
PiPrius:WA Manzanita Micro, PiPrius, AVI, Green Car Co.	No	Yes	Unknown	4	10	20-30	90 to 300 Vac Vdc	0.4-3+	Flame Spill proof	150	no	$10K +Labor	0	Hyb DC CV	4.7	4+0.3 stock	PbA (Ni, Li)
EnergyCS:CA EnergyCS	Yes	No	Unknown	11	30	50	120 Vac	9.0	Flame Spill proof	83	no	$40K	0	New	9	8	Li
Amberjac:UK Amberjac EnergyCS partner	Yes	No	Dev	7	30	60-70	110 to 230 Vac	9.0	Flame Spill proof	83	yes	$40K	0	New	9	8	Li
EDrive:CA EDrive Systems	?	?	Unknown	0	32?	60?	100 to 240 Vac	9.0	via cell sep	?	yes	$12K	0	New	9.5	8.5?	Li
Hymotion:ON Canada Hymotion / A123 (PHEV-L5)	Yes, fleets	No	Prod	18	15	30	100 to 240 Vac	5.5 / 4.0	Spill proof	72	no	$10K	3	Hyb	5.0	4+0.3 stock	Li
Plug-In Conversions:CA Plug-In Conversions	Yes	Yes	Prod	15	25	50	120/240 Vac	6/2	Flame Spill proof	100	no	$12.5K	3	New CV	6.1	5.1	NiMH
OEMtek:CA OEMtek	Yes	No	Unknown	?	30	50	100/240 Vac	4/6	Flameproof	95	Yes	$12K	0	Hyb	9	8	Li
3Prong Power:CA 3Prong Power	Yes	No	Prod	12	10-12	20+	110 Vac	4	Flame Spill proof	130	Yes	$6.7K	1	Hyb	4.8	2.4	PbA

Plug-In Supply of Petaluma is creating conversion kits that have all of the necessary components already assembled. Everybody agrees that the conversion process isn't cheap. But the price of oil—including greenhouse gas emissions and war—makes plug-ins an increasingly attractive option, at least until the car companies get in gear.

"Had it not been for the grassroots effort," Sherwood said, "backyard conversions wouldn't be possible. Car companies wouldn't even be thinking about making plug-in hybrids." But they're thinking about it now.[5]

PHEV Conversion Decisions

When you do a PHEV conversion today, you have many chassis choices: small car, sports car, compact car, crossover, SUV (not recommended), or small truck. Small cars and sports cars may weigh less, but they also have less room for batteries and minimal payload. Most vehicle models gain 25 to 50 pounds each year as the manufacturers add more auxiliaries or sound-deadening materials. Cars and crossovers have the advantage of less aerodynamic drag, an advantage that SUVs and trucks do not have.

Whatever vehicle you choose, select one that you personally like. Why spend $10,000 and approximately 100 hours on a vehicle that you do not like? This is a vehicle you want to show proudly.

You have additional choices of an AC (alternating current) or DC (direct current) drive system. Typically, AC systems use higher voltages and give faster performance, but they may cost two or three times more than a simpler and proven DC system. Each system requires an ICE motor, an electric motor, controllers, batteries, and a charger. This section will look at these choices and prepare you for the guidance given by the rest of the book.

To make the best decisions, you must first identify your requirements. Are you looking for a small commuter vehicle for yourself, or do you need an in-town vehicle for a small family? The number of people, performance, and range are all basic considerations. The cost of the drive system is directly related to your requirements.

Therefore, if you are on a limited budget, it is critical that you distinguish between "requirements" and "desires."

Your Chassis Makes a Difference

If you're going ahead with the conversion alternative, your most important choice is the chassis you select. Use the Internet to find the curb weight and payload capacity of the chassis you're considering. Note the manufacturer's specs on the driver's door when you look at a vehicle. Once you've identified a few potential candidate vehicles, make cardboard mock-ups of the batteries you might use. Then find a vehicle and see if they will fit.

Minimizing weight is always the number one objective of any PHEV conversion. When this is added to the criteria of minimizing conversion time and maximizing the odds of get-it-right-the-first-time success, the trade-off points squarely in the direction of the pickup truck. The van weighs more, and the car is typically the most time-consuming conversion (the smaller amount of room in which to mount PHEV options increases the problem of getting parts to fit).

The internal combustion vehicle Toyota Prius is actually an outstanding plug-in hybrid electric vehicle conversion choice because you need only extra battery packs and an additional controller to communicate with the existing vehicle. This is true not only of the Prius, but also of the Ford Escape and any other hybrid on the market.

Your Batteries Make a Difference

Of all the advantages of a pickup truck, its extra room makes the biggest difference because it means that there is more space available to mount batteries. With car conversions, you must either choose a larger chassis or go to 8- or 12-volt batteries. A Toyota PRIUS+ conversion retains the OEM hybrid battery and its management computer while adding a lead-acid pack consisting of 20 BB Battery EVP20-12B 12-volt, 20-ampere-hour sealed AGM lead-acid batteries. Lead-acid battery chemistry is very inexpensive, but it has significant limitations. However, with this relatively inexpensive conversion (as little as $4,000 in parts cost, including the battery), you can be the first in your community to actually own and drive a plug-in hybrid, and you can achieve 100+ mpg (plus electricity) for 15 to 20 miles/day![6]

More advanced batteries may be able to be retrofitted to the conversion. Any new battery's enclosure, mounting, and thermal management system will no doubt also be very different. See Table 4-2 for the PRIUS+ costs using lead-acid versus

Table 4-2 Battery Choices with Cost, EV Mileage, and Weight
Source: http://www.eaa-phev.org/wiki/PriusPlus.

Chemistry		Usable Wh/kg	Cycle life	Yr daily driving	$/usable kWh	$/kWh thruput	Cents/EV-mi	kWh	$	EV mi	Wt, lb
PbA (current)		16	400	1.1	$380	$0.95	20.0	2.1	$798	10	289
NiMH	Worst	36	2000	5.5	$1,200	$0.60	12.6	4.2	$5,040	20	257
NiMH	Best	36	4000	11.0	$800	$0.20	4.2	4.2	$3,360	20	257
Li-ion	Worst	56	1000	2.7	$1,200	$1.20	25.2	4.2	$5,040	20	165
Li-ion	Best	100	4000	11.0	$800	$0.20	4.2	6.3	$5,040	30	139
NiZn	Worst	36	500	1.4	$500	$1.00	21.0	4.2	$2,100	20	257
NiZn	Best	36	2000	5.5	$350	$0.18	3.7	4.2	$1,470	20	257
Firefly PbA	Worst	36	1000	2.7	$350	$0.35	7.4	4.2	$1,470	20	257
Firefly PbA	Best	45	4000	11.0	$250	$0.06	1.3	5.25	$1,313	25	257

lithium ion batteries and how the choice affects energy cost and the life, range, and weight of the vehicle.

Table 4-2 presents all the dimensions of the conversion trade-off with regards to batteries—rated as best and worst. This chart shows you don't need money to be an obstacle. If you do not have a budget for a PHEV, you can customize your PHEV; add the latest, most powerful batteries with the sole purpose to tool around the countryside powered by nickel–metal hydride or lithium polymer batteries. We all want fully electric but the rest of us have to take it a bit slower. The movement toward the best battery can proceed only as fast as our pocketbooks allow: readily available batteries with high energy density will only get these prices better. However 1.3 cents per mile works for me!

The Procedure

Chapters 5 through 9 introduce you to chassis, motors, controllers, batteries, and chargers, and Chapter 12 provides you with some sources to get you started. In Chapter 10, you'll look over my shoulder while I convert a Ford Ranger, following step-by step instructions that you can adapt to nearly any conversion you want. Chapter 11 shows you how to maximize the enjoyment of your PHEV once it's up and running. Use the sources in Chapter 12—don't just take my word for it. Join the EAA-PHEV Group and the EAA. Subscribe to their newsletter. Read all the books, magazines, and research material you can. There is also a lot of useful material online (in other words, surf the Net!). Go to meetings, shows, and rallies. Most of all, talk to people who have already done a conversion. If you listen to what they say, you will soon discover there are more opinions on how to do an electric vehicle conversion than there are snowflakes in the known universe. Then integrate all this information and make your own decisions. After you've done your first conversion, you'll notice a new phenomenon—people will start listening to you.

How Much Is This Going to Cost?

Notice also that the professionals tell you what performance you can expect, when you're going to get it, how much it's going to cost, and for how long the quoted prices are valid. Be sure you get the same information, in writing, from any supplier you choose. It is important that you select components that will perform as a system. Don't expect to buy random components from the Internet and then get them to function properly as a system. In addition, professionals should be available after the sale to assist you with any problems.

Table 4-3 shows the prices for all parts associated with such conversions. Based on range and cost, the table shows indirectly what extra costs to expect when using the latest new batteries and chassis and a few extra bells and whistles. The amounts that you might be able to obtain from selling off the internal combustion engine components were omitted from the comparisons; you can expect the vehicle costs to be lower if you do sell them.

TABLE 4-3 Costs Associated with a PRIUS+ Conversion

Estimated Component Costs	Minimum	Maximum
Battery set (20 + 2 spares)	$900	$1100
Battery wire & lugs	$100	$150
Heating pads & insulation	$100	$100
CAN-View	$600	$600
Display (opt for 2004-5 Prii)	$0	$200
Charger (Delta-q or Brusa)	$800	$2500
Cord reel & base, brackets	$100	$100
Contactors (3)	$240	$330
Fuses & holders 60 A (2)	$100	$150
Fans (3)	$60	$120
All metal & plastic	$200	$300
Circuit board	$100	$100
Circuit board components	$200	$300
Connectors	$200	$300
Misc. electronics	$150	$200
Total	$3850	$6550
Estimated Fabrication Costs		
Assembled and tested circuit board	$250–500	
Battery tray (4 needed)	$150–250	
Battery box top	$150–200	
Battery box foundation	$150–200	
Electronics tray, assembled and wired	$500–1000	
Set of pre-built battery cables	$150–200	
Pre-built low-power wiring harness	$150–300	
Total (including 4 trays)	$1950–2500	

CHAPTER 5

Power Trains and Designing Your PHEV

To understand how to build your own PHEV, you need to understand the different types of drive systems for hybrid electric vehicles and how they work. Once we have gone through how a hybrid works and the types of hybrid electric cars, then we can discuss how to design your PHEV for conversion.[1]

How Does That Hybrid Electric Car Work in the First Place?

Hybrid electric vehicles combine the benefits of gasoline engines and those of electric motors to provide improved fuel economy.[2]

The gasoline engine provides most of the vehicle's power, and the electric motor provides additional power when needed, such as for accelerating and passing. This allows a smaller, more efficient engine to be used.

The electric power for the motor is generated from regenerative braking and from the gasoline engine, so hybrids don't have to be plugged into an electrical outlet to recharge.

Types of Hybrid Drive Systems

Series Hybrids

A hybrid electric vehicle (HEV) with a series configuration[3] uses the heat engine or fuel cell with a generator to produce electricity for the battery pack and the electric motor. Series HEVs have no mechanical connection between the hybrid power unit and the wheels; this means that all motive power is transferred from chemical energy to mechanical energy to electric energy, then back to mechanical energy to drive the wheels. Here are some benefits of a series configuration:

- The engine never idles, which reduces vehicle emissions.
- The engine can operate continuously in its most efficient region.
- The engine drives a generator to run at optimum performance.
- The design allows for a variety of options when mounting the engine and vehicle components.
- Some series hybrids do not need a transmission.

Downside to Series HEVs

The downside is that series HEVs require larger, and therefore heavier, battery packs than parallel vehicles. In addition, the engine has to work hard to maintain the battery charge because the system is not operating in parallel. There is also the inefficiency of converting the chemical energy to mechanical energy to electric energy and back to mechanical energy.

Series hybrids have also been called extended-range electric vehicles, since they use an internal combustion engine (ICE) to turn a generator. That generator provides current to the electric motor, and the motor drives the wheels.

An example of a series hybrid is the Renault Kangoo Elect'Road, shown in Figure 5-1. Figure 5-2 shows a typical layout for a series hybrid.

Figure 5-1 The Renault Kangoo is a series hybrid electric vehicle.

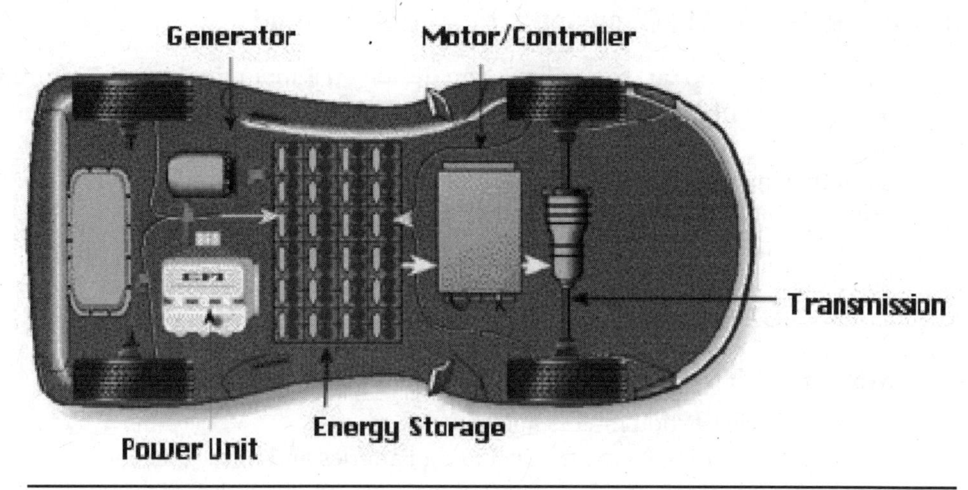

Figure 5-2 Layout of a series hybrid electric vehicle drive system. Source: http://www.afdc.energy.gov/afdc/vehicles/hybrid_electric_series.html.

Hybrid Electric Vehicle Parallel Design

A hybrid electric vehicle (HEV) with a parallel configuration[4] has a direct mechanical connection between the hybrid power unit and the wheels, as in a conventional vehicle, but it also has an electric motor that drives the wheels. For example, a parallel vehicle could use the power created by an internal combustion engine for highway driving while using power from both the engine and the electric motor for accelerating. Some benefits of a parallel configuration are

- A smaller engine provides more efficient operation and therefore better fuel economy without sacrificing acceleration power. The vehicle has more power because both the engine and the motor supply power simultaneously.
- Most parallel vehicles do not need a separate generator because the motor regenerates the batteries.
- Power does not need to be redirected through the batteries and can therefore be more efficient.

Honda uses the parallel hybrid drive system on its Insight, Civic, and Accord.

Series-Parallel

The hybrid power trains currently used by Ford, Lexus, Nissan, and Toyota, which some refer to as "series-parallel with power split," can operate in both series and parallel modes at the same time.[5]

Designing Your PHEV: Choosing a Mounting Method

Before you get too far, you will need to decide on a mounting method. Currently, there are two methods, each with its benefits. The majority of the instructions are the same for both methods; however, some sections will be labeled as "classic mounting method" and others as "alternative mounting method" and contain instructions for the specific method.

A hybrid electric vehicle employs an optimized mix of various components. You can see a typical hybrid configuration in Figure 5-2 and learn more about the various HEV components from the links given in the notes.

Hybrid Electric Vehicle Drivetrain Components and System Details[6]

- Electric traction motors and controllers
- Energy storage systems, including batteries and ultracapacitors
- Power units and transmissions, including spark ignition engines, compression-ignition and direct-injection (diesel) engines, gas turbines, and fuel cells
- Energy management and systems control

How to Make and Design That PHEV

Figure 5-3 shows the typical layout of a hybrid electric vehicle. The funny part is that it is simply an electric car with a gasoline engine. Figure 5-4 shows the layout of the PHEV, which includes a charger and extra batteries for more range. That is why it is *so important* that you understand an electric vehicle first, then an internal combustion engine (ICE). Why? Everyone pretty much understands how an ICE works. Understanding the electric car as well makes it easy to figure out how they work together to make a great car.

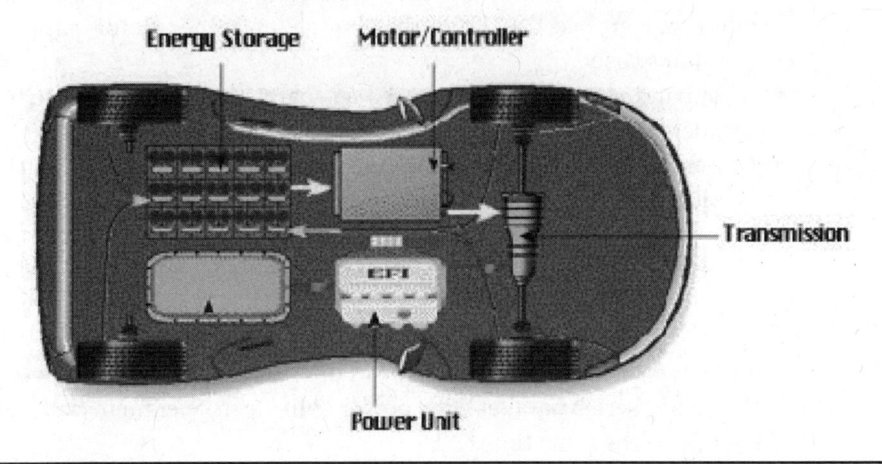

FIGURE 5-3 Layout of a hybrid electric vehicle. Source: U.S. Department of Energy.

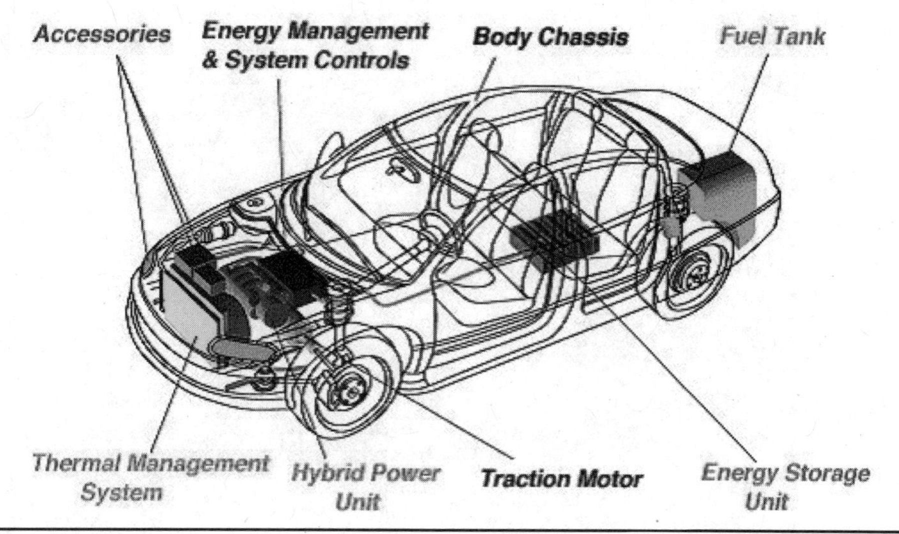

Figure 5-4 Layout of a plug-in hybrid electric vehicle. Source: U.S. Department of Energy.

Classic Mounting Method

The "classic" method is the style that CalCars uses. The spare tire well is largely unchanged, the batteries are placed in the very rear of the trunk, an electronics tray is placed in front of the batteries, and the charger is placed in the left cubbyhole (the carpeting in the cubbyhole is cut back). This method allows easy access to all the electronics, including the charger. Figures 5-5 to 5-8 show different elements of the classic mounting method.

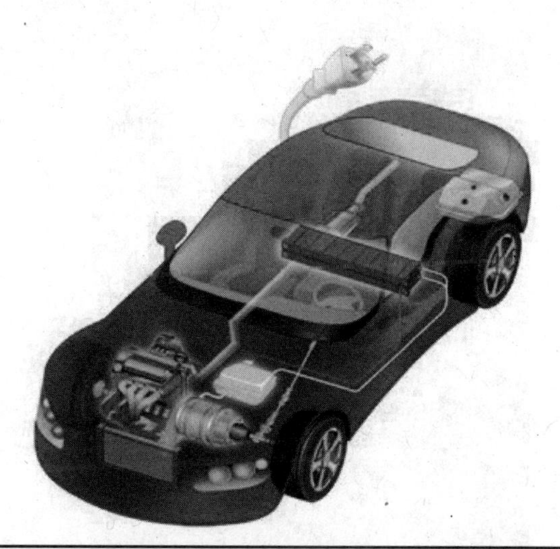

Figure 5-5 The PRIUS+ conversion of Sven's Prius from November 2006. Classic method, showing the spare tire well, 110-volt breakout quad, and 110- to 120-volt power supply.

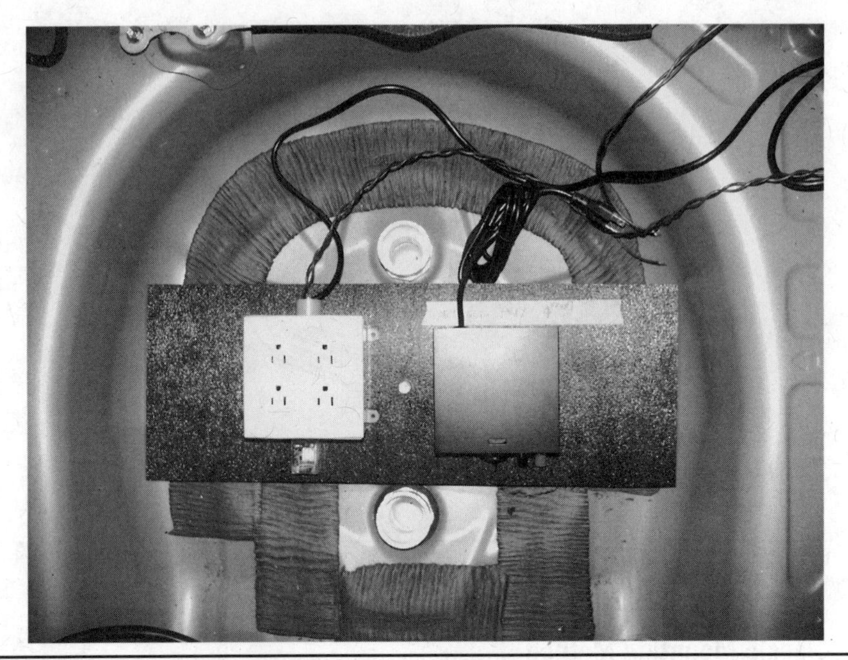

Figure 5-6 Classic method: spare tire well.

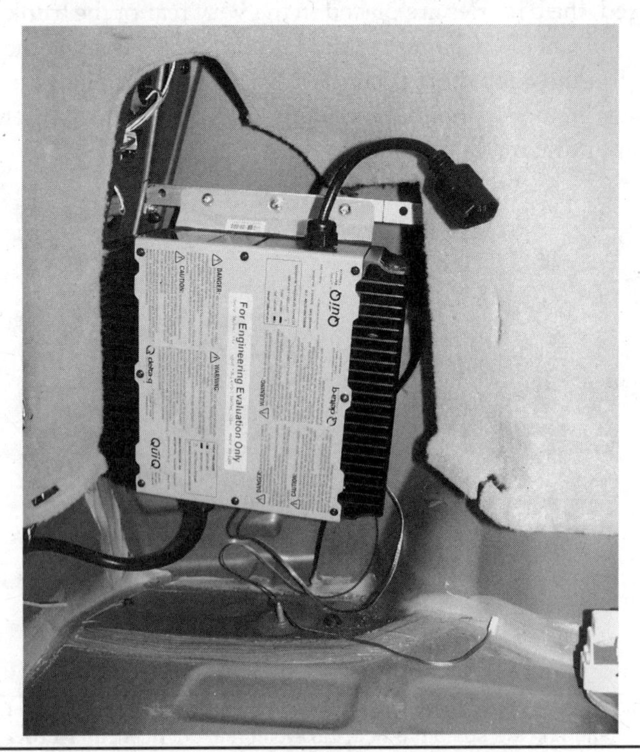

Figure 5-7 Classic method: Delta-Q charger installed.

Figure 5-8 Finished trunk (top view). Finished view (minus Plexiglas) of the classic mounting method—slightly outdated now, but pretty much accurate.

Alternative Mounting Method

The "alternative" method seeks to maximize usable trunk space and minimize visible changes. With this design, the batteries are moved as far forward as possible, and the supporting electronics and charger are located in the spare tire well. The batteries stick up through the false floor, but are flush with the Prius's floor. Most of the high-voltage electronics are isolated from the low-voltage electronics; however, some of the high-voltage electronics and the charger are more difficult to access because they are below the batteries (they tend to be the more reliable parts, however). The low-voltage electronics are easily accessible. Figures 5-9 to 5-11 show different aspects of the alternative mounting method.

- Two models: Prius-15 and Prius-30
- Pure EV range (<34 mph): 15 miles/30 miles
- PHEV range (conservative driving, after full charge): 25 miles/50 miles
- PHEV fuel efficiency (conservative, full charge): 100 mpg
- Conversion locations:
  - In place of the OEM battery, and taking some of the space occupied by the black tray in the trunk

FIGURE 5-9 Spare tire well. Source: http://www.eaa-phev.org/images/c/c1/Electronics_Box_And_Charger_in_Tire_Well.jpg.

FIGURE 5-10 Location of batteries. Source: http://www.eaa-phev.org/images/b/bc/Batteries_being_installed_2.jpg.

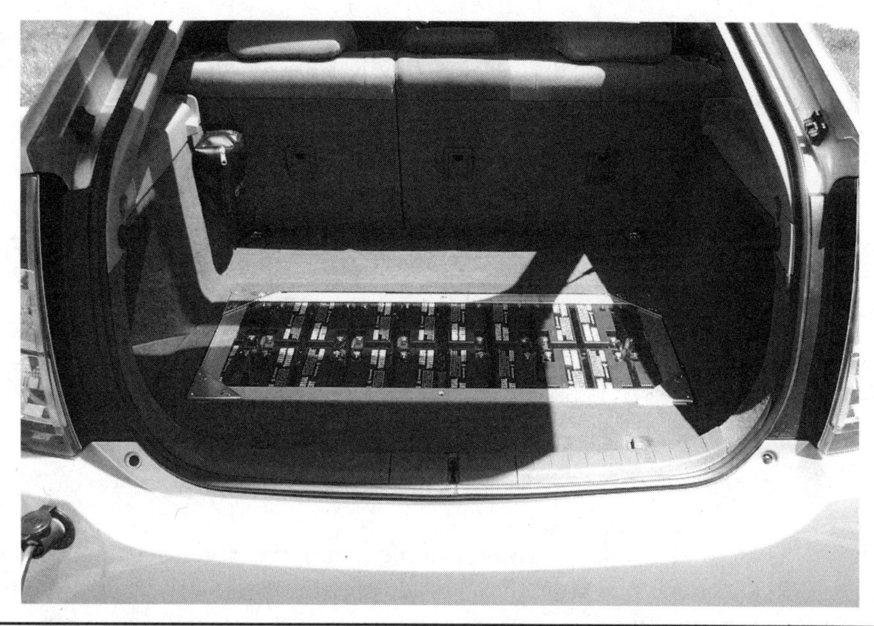

Figure 5-11 Finished product. Source: http://www.eaa-phev.org/wiki/Image:Finished_Alt_Mounting_Method.jpg.

- Maintains access to the spare tire, in its original location
- Charge plug in rear bumper

The chassis is the foundation of your electric vehicle conversion. While you might never build your own chassis from scratch, there are fundamental chassis principles that can help you with any EV conversion or purchase—things that never come up with internal combustion engine vehicles—such as the influence of weight, aerodynamic drag, rolling resistance, and drivetrains.

This chapter will step you through the process of optimizing, designing, and buying your own electric vehicle. You'll become familiar with the chassis trade-offs involved in optimizing your EV conversion. Then you'll design your EV conversion to be sure the components you've selected accomplish what you want to do. When you have figured out what's important to you and verified that your design will do what you want, you'll look at the process of buying your chassis.

Drivetrains

Let's start with what the drivetrain[7] in a conventional internal combustion engine vehicle must accomplish. In practical terms, the power available from the engine must be equal to the job of overcoming the tractive resistances discussed earlier for any given speed.

The obvious mission of the drivetrain is to apply the engine's power to driving the wheels and tires with the least loss (highest efficiency). But overall, the drivetrain must perform a number of tasks:

- Convert the torque and speed of the engine to vehicle motion and traction
- Change directions, enabling forward and backward vehicle motion
- Permit different rotational speeds of the drive wheels when cornering
- Overcome hills and grades
- Maximize fuel economy

There is a standard drivetrain layout that is most widely used to accomplish these objectives today. The function of each component is as follows:

- *Engine (or electric motor).* This provides the raw power to propel the vehicle.
- *Clutch.* For internal combustion engines, the clutch separates or interrupts the power flow from the engine so that the transmission gears can be shifted and, once the gears are engaged, the vehicle can be driven from a standstill to top speed.
- *Manual transmission.* This provides a number of alternative gear ratios to the engine so that a wide range of vehicle needs—from maximum torque for hill climbing or minimum speed to economical cruising at maximum speed—can be accommodated.
- *Driveshaft.* A device that connects the drive wheels to the transmission in rear-wheel-drive vehicles; it is not needed in front-wheel-drive vehicles.
- *Differential.* A device that accommodates the fact that the outer wheels must cover a greater distance than the inner wheels when a vehicle is cornering, and that translates the drive force 90 degrees in rear-wheel-drive vehicles (it might or might not do so in front-wheel-drive vehicles, depending on how the engine is mounted). Most differentials also provide a speed reduction with a corresponding increase in torque.
- *Drive axles.* These transfer power from the differential to the drive wheels.

You can typically expect 90 percent or greater efficiencies (slightly better for front-wheel-drive vehicles) from today's drivetrains. Internal combustion engine vehicle drivetrains provide everything necessary to allow an electric motor to be used to propel the vehicle in place of the removed internal combustion engine and its related components. However, the drivetrain components are usually complete overkill for the EV owner. The reason has to do with the different characteristics of internal combustion engines and electric motors, and the way they are specified.

Difference in Motor vs. Engine Specifications

Comparing electric motors and internal combustion engines is not an "apples to apples" comparison. If someone offers you either an electric motor or an internal combustion engine with the same rated horsepower, take the electric motor—it's far more powerful.

Also, a series-wound electric motor delivers peak torque upon start-up (zero rpm), whereas an internal combustion engine delivers nothing until you wind up its rpm.

An electric motor is so different from an internal combustion engine that a brief discussion of terms is necessary before going further.

There is a substantial difference in the way an electric motor and an internal combustion engine are rated in horsepower. The purpose of Figure 5-12 is to show at a glance that an electric motor is more powerful than an internal combustion engine with the same rated horsepower. All internal combustion engines are rated at specific rpm levels for maximum torque and maximum horsepower. Maximum horsepower ratings for internal combustion engines are typically derived under idealized laboratory conditions (for the bare engine without accessories attached), which is why the rated horsepower point appears to be above the maximum peak of the internal combustion engine horsepower curve in Figure 5-12. Electric motors, on the other hand, are typically rated at the continuous output level they can maintain without overheating. As you can see from Figure 5-12, the rated horsepower point for an electric motor is far down from its short-term output, which is typically two to three times higher than its continuous output.

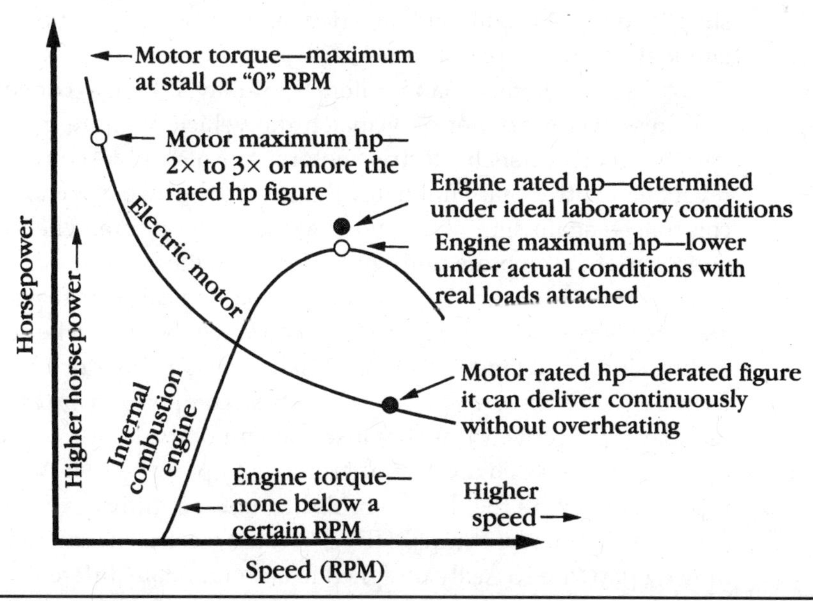

Figure 5-12 Comparison of characteristics of the electric motor and the internal combustion engine.

There is another substantial difference: while an electric motor can produce a high torque at zero speed, an internal combustion engine produces negative torque until some speed is reached. An electric motor can therefore be attached directly to the drive wheels and accelerate the vehicle from a standstill without the need for the clutch, transmission, or torque converter required by the internal combustion engine.

Everything can be accomplished by controlling the drive current to the electric motor. While an internal combustion engine can deliver peak torque only in a relatively narrow speed range and requires a transmission and different gear ratios to deliver its power over a wide range of vehicle speeds, an electric motor can be designed to deliver its power over a broad speed range with no need for a transmission at all.

All these factors mean that current EV conversions put a lighter load on drivetrains borrowed from an internal combustion engine vehicle, and future EV conversions will eliminate the need for several drivetrain components altogether. Let's briefly summarize:

- *Clutch.* Although it is basically unused, a clutch is handy to have in today's EV conversions because its front end gives you an easy place to attach the electric motor, and its back end is already conveniently mated to the transmission. In short, it saves the work of building adapters and other such devices. In the future, when widespread adoption of AC motors and controllers will eliminate the need for a complicated mechanical transmission, it will be possible to directly couple the electric motor to a simplified, lightweight, one-direction, one- or two-gear-ratio transmission, eliminating the need for a clutch.
- *Transmission.* Another handy item in today's PHEV conversions, the transmission's gears not only match the vehicle you are converting to a variety of off-the-shelf electric motors, but also give you a mechanical reversing control that eliminates the need for a two-direction motor and controller—again simplifying your work. In the future, when widespread adoption of AC motors and controllers provides directional control and eliminates the need for a large number of mechanical gears to get the torques and speeds you need, today's transmission will be able to be replaced by a greatly simplified (and even more reliable) mechanical device.
- *Driveshaft, differential, and drive axles.* These components are all used intact in today's EV conversions. Because contemporary, built-from-the-ground-up electric vehicles like General Motors's Impact use two AC motors and place them next to the drive wheels, it's not too difficult to envision even simpler solutions for future PHEVs, because electric motors (with only one moving part) are so easily designed to accommodate different roles.

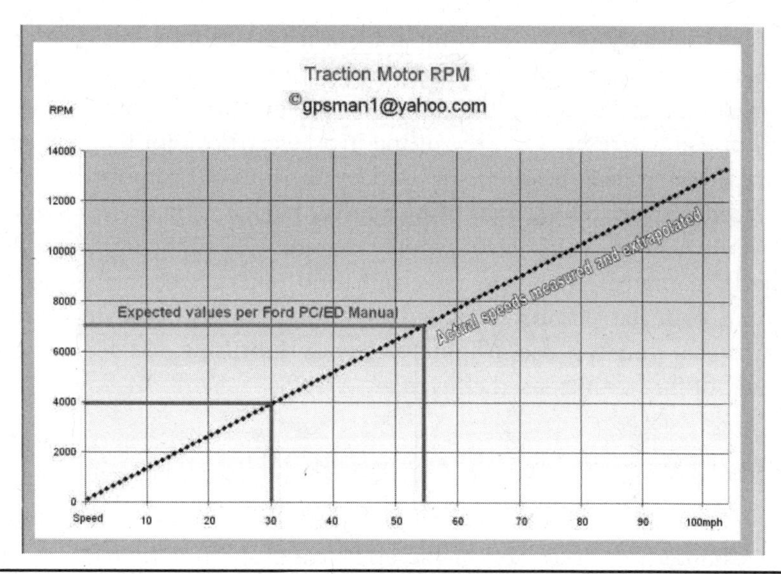

FIGURE 5-13 Traction motor rpm. Source: EAA-PHEV website.

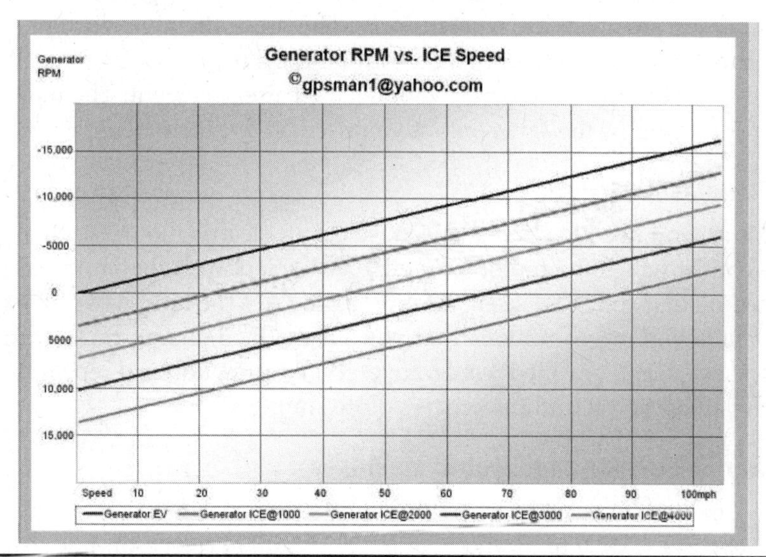

FIGURE 5-14 Generator rpm vs. ICE speed on a Ford Escape PHEV. Source: EAA-PHEV website. Source: RPM chart and data courtesy of gpsman1@yahoo.com and the EAA-PHEV.

Notes: Traction motor:generator ratio = 1:1.21875 when in EV mode. Each 1,000 engine rpm raises generator speed by 3,394 rpm.

Traction motor rpm is always relative to wheel speed.

Traction motor spec is 68 kW (91 hp) in Ford's Service Manual (70 kW/94 hp in the consumer brochure).

Generator motor spec is 28 kW (37 hp) in Ford's Service Manual (30 kW/40 hp in the consumer brochure).

The corollary of all this is the lubricant you choose. Using a lighter-viscosity fluid in your differential lets things turn a lot more easily. You're not breaking any rules here. Instead of shoving 500 horsepower through your drivetrain, you're at the opposite extreme—you're putting in an electric motor that lets you cruise at 10 percent of the peak torque load used by the internal combustion engine you just replaced. You're shifting less, using a lower peak torque, and probably using it less often. As a result, your electric motor is putting only the lightest of loads on your internal combustion vehicle drivetrain, and you're probably using 50 percent (or less) of your drivetrain's designed capability. So less wear and tear on the gears means that you can use a lighter-viscosity lubricant and recover the additional benefit of further increased efficiency.

Design Your EV

Another point (covered in more detail in Chapter 6's discussion of electric motors and Chapter 9's discussion of the electrical system) is that you need to think in terms of current when working with electric motors. The current is directly related to motor torque. Through the torque-current relationship, you can directly link the mechanical and electrical worlds. (*Note:* The controller gives current multiplication. In other words, if the motor voltage is one-third the battery voltage, then the motor current is slightly less than three times the battery current. The motor and battery current would be the same only if you used a very inefficient resistive controller.)

Calculation Overview

In short, you need to select a speed, select an electric motor for that speed, choose the rpm at which the motor delivers that horsepower, choose the target gear ratio based on that rpm, and see if the motor provides the torque over the range of level and hill-climbing conditions that you need. Once you go through the equations, worksheets, and graphed results covered in this section and repeat them with your own values, you'll find the process quite simple.

Power Electronics and Electric Machines

Drive motors/generators in today's hybrids are packaged as fully integrated front-wheel-drive (FWD) units, like that in the original Prius; as in-line rear-wheel-drive (RWD) units, such as that in the 2007 Lexus LS 600h; or as axle-mounted RWD units, such as that in the Lexus RX 400h.

Power ranges from 50 kW maximum (at 1,200–1,540 rpm for the approximately 25 kW continuous Prius motor) up to 160 kW maximum for the Lexus LS 600h. But in all cases, the electric traction motors provide about half the maximum power of their respective propulsion systems.

High-power electric rear drive, such as in the "two-mode" system being developed by the joint venture of GM, BMW, and DCX, appears to be the preferred design direction in the premium hybrid market, providing four-wheel drive (4WD) to boost performance in the LS motor, generator, power split planetary gear mechanism, and speed reduction in one transmission casing. Saab reaches for maximum performance in the BioPower hybrid concept vehicle, which utilizes both the FWD version of the two-mode system and an electric rear axle.

Power electronics are designed to match the 30-kW, 244–650-volt DC/DC boost converter to the characteristics of the battery and traction motor.

The 2007 Toyota Camry 105-kW motor inverter/75-kW generator inverter integrated power unit exemplifies the state of the art: a 2,098-uF, 750-V DC, 15.4-kg package capacitor bank that replaces the standard starting battery and contains the traction drive.

Parasitic Loss Reduction[8]

Heavy vehicles lose a tremendous amount of energy to wind resistance and drag, braking, and rolling resistance. Such nonengine losses can account for an approximately 45 percent decrease in efficiency. Other sources of energy loss include friction and wear in the power train, thermal (heat) loads, operation of auxiliary loads (air conditioning, heaters, refrigeration, and so on), and energy lost by trucks when their engines continue to run while parked, a practice known as idling.

The parasitic loss reduction activity identifies methodologies that may reduce energy losses and tests them in the laboratory. Promising technologies are then prototyped and tested onboard heavy vehicles. Once validated, the technologies must be tested on the road to obtain durability, reliability, and life-cycle cost data for the developmental component and/or design strategy.

Current areas of focus for the parasitic loss reduction activity include

- Aerodynamic drag reduction, to characterize and respond to energy losses caused by wind and rolling resistance
- Friction and wear reduction, to understand and address the multiple surface interactions that occur in heavy vehicle systems
- Regenerative shock absorbers primarily for trains, which can recover energy that is dissipated by conventional shock absorbers
- Predictive cruise control that can control vehicle speed for optimal fuel efficiency
- Idle reduction devices and systems that enable truckers to turn off their engines when stopped and still be comfortable while sleeping
- Locomotive systems that reduce emissions and increase fuel efficiency
- Off-highway systems for construction and farm vehicles
- Diesel reformers to transform diesel fuel into carbon monoxide and hydrogen for use with auxiliary systems
- Thermal management to counter some of the negative heat-producing consequences of emissions control techniques

From the figures and curves listed in Table 5-1, you can derive the rpm at which your electric motor delivers closest to its rated horsepower.

TABLE 5-1 RPM, Electric Motor, and Horsepower Matrix

MPH	Engine RPM	Traction Motor RPM	Generator Motor RPM
1	0 (EV)	128	–156
2	0 (EV)	256	–312
5	0 (EV)	640	–780
10	0 (EV)	1280	–1560
40	0 (EV)	5120	–6240
40	1000	5120	–2846
40	2000	5120	548
40	3000	5120	3942
40	4000	5120	7336
60	2000	7680	–2572
60	4000	7680	4216
80	2000	10,240	–5692
80	4000	10,240	1096

Table 5-2 also shows you how to design the vehicle using the PRIUS+ design with the Toyota Prius.

TABLE 5-2 Design That Prius as a PRIUS+

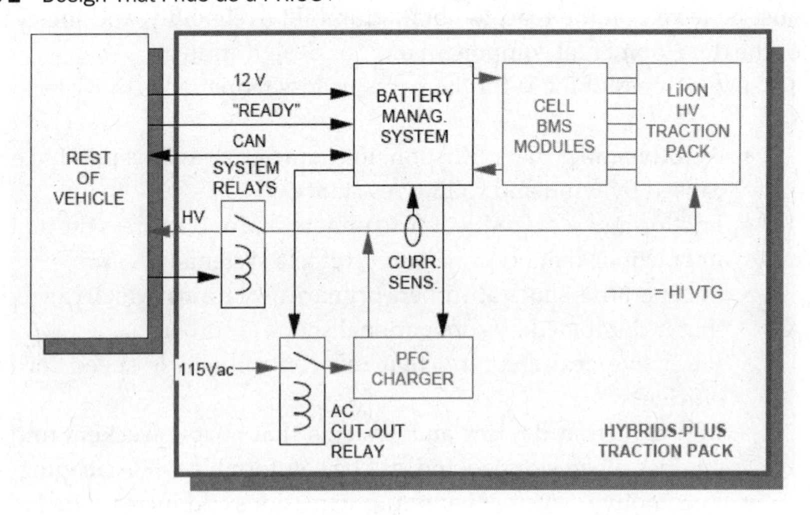

Buy Your EV Chassis

Even if you go out to buy your PHEV ready-made, you still want to know what kind of a job the manufacturer has done, so that you can decide whether you're getting the best model for you. In all other cases, you'll be doing the optimizing,

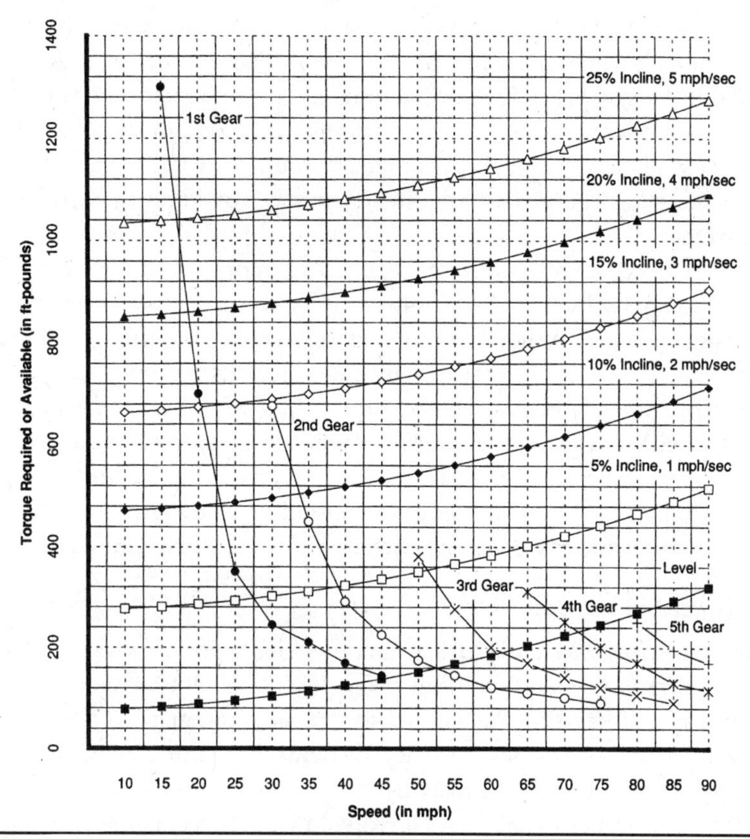

FIGURE 5-15 Electric vehicle torque required versus torque available.

either by the choices you've made up front in chassis selection or by other decisions that you make later on during the conversion. In this section, you'll be looking at key points that contribute to buying smart:

- Review why conversion is best—the pro side.
- Consider why conversion might not be for you—the con side.
- Find out how to get the best deal.
- Keep your needs list handy.
- Buy or borrow the chassis manual.

Why Conversion Is Best

In the real world, where time is money, converting an existing internal combustion engine vehicle rather than building a new PHEV saves money in that it avoids a large capital investment and a large amount of labor. The bonus for the PHEV converter when starting with an existing late-model hybrid vehicle is a structure that comes complete with body, chassis, suspension, steering, and braking systems—all designed, developed, tested, and safety-proven to work together. As long as the converted electric vehicle does not greatly exceed the original vehicle's Gross Vehicle Weight Rating (GVWR) overall weight or Gross Axle Weight Rating (GAWR) weight per axle specifications, all systems will continue to deliver their previous performance, stability, and handling characteristics. And the PHEV converter inherits another body bonus: the bumpers, lights, safety-glazed windows, and other such equipment are already preapproved and tested to meet all safety requirements.

There's still another benefit—you save more money. Automobile junkyards make money by buying the whole car (truck, van, or whatever) and selling off its pieces for more than they paid for the car. When you build (rather than convert) a PHEV, you are on the other side of the fence. Unless you buy a complete kit, building from scratch means buying chassis tubing, angle braces, and sheet stock plus axles/suspension, brakes, steering, bearings/wheels/tires, body/trim/paint, windshield/glass/wipers, lights/electrical, gauges, instruments, dashboard/interior trim/upholstery, and so on—parts that are bound to cost you more à la carte than when you buy them already manufactured and installed in a completed vehicle.

The Other Side of Conversion

What's the downside? It's likely that any vehicle you choose to convert will not be streamlined like a soapbox derby racer. It will be a lot heavier than you'd like it to be, and it will have tires designed for traction rather than low rolling resistance. You do the best you can in these departments depending on your end-use goals: EV dragster, commuter, or highway flyer.

It's equally likely that your conversion vehicle will come with a lot of parts that you no longer need: the internal combustion engine and mounts, and its fuel, exhaust, emission control, ignition, starter, and cooling/heating systems. These you remove and, if possible, sell. Then there are additional components that you might wish to change or upgrade for better performance, such as the drivetrain, wheels/tires, brakes, steering, and battery/low-voltage accessory electrical system. On these, just do what makes sense.

Keep Your Needs List Handy

Regardless of which vehicle you choose for conversion, you want to feel good about your ability to convert it before you leave the lot. If it's too small and/or cramped to fit all the electrical parts—let alone the batteries—you know you have a problem. Or if it's very dirty, greasy, or rusty, you might want to think twice. Here's a short checklist to keep in mind when buying:

- *Weight*. With 120 volts and a 22-hp series DC motor, 4,000 to 5,000 pounds is about the upper limit. On the other hand, the same components will give you blistering performance and substantially more range in a 2,000- to 3,000-pound vehicle. Weight is everything in EVs—decide carefully.

- *Aerodynamic drag*. You can tweak the nose and tail of your vehicle to produce less drag and/or turbulence, but what you see before you buy is basically what you've got. Choose wisely and aerodynamically.
- *Rolling resistance*. Special EV tires are still expensive, so look for a nice set of used radials and pump them up hard.
- *Drivetrain*. You don't want an automatic; a four- or five-speed manual will do nicely, and front-wheel drive typically gives you more room for mounting batteries. Avoid eight and six cylinders in favor of four cylinders, and choose the smallest, lightest engine/drivetrain combinations. Avoid heavy-duty anything or four-wheel drive.
- *Electrical system*. Pass on air conditioning, electric windows, and any power accessories.
- *Size*. Will there be room for everything you want to put in (batteries, motor, controller, and charger)? How easy will it be to do the wiring?
- *Age and condition*. These determine whether you can get parts for the vehicle and how easy it is to restore it to a condition where it's fit to serve as your car.

Buy or Borrow the Manuals

Manuals are invaluable. If possible, seek them out so that you can read about any hidden problems before you buy the vehicle. After you own it, don't spend hours figuring out whether the red-striped or the green-striped wire goes to dashboard terminal block number 3; just flip to the appropriate schematic in the manual and locate it in minutes. Component disassembly is easy when you know that you must always disengage bolt number 2 in a clockwise direction before turning bolt number 1 in a counterclockwise direction and other such information. Believe me, these are labor savers.

Design That Prius as a PHEV!

Table 5-2 shows you a design for the PRIUS+. If you like that design, work with the PRIUS+ design and get assistance. Figure 5-16 shows the original battery pack installation for the Prius, and Figure 5-17 shows the PHEV battery pack location.

More importantly, if you are still confused and need help, the people at the Plug In Center (Figure 5-18) are there to help. Don't be afraid; just *do it*!

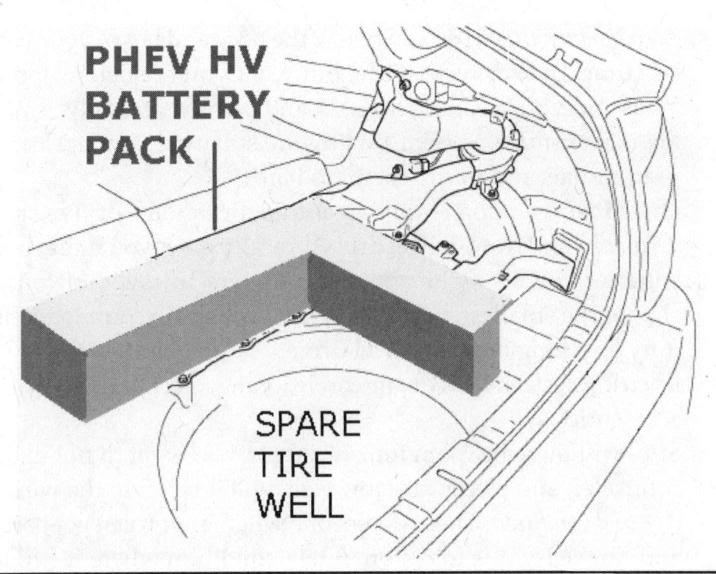

FIGURE 5-16 Prius original battery pack. Sources: EAA-PHEV; PRIUS+.

FIGURE 5-17 The PHEV battery pack location. Sources: EAA-PHEV; PRIUS+.

Figure 5-18 The Plug In Center Crew getting busy with a PHEV conversion (www.plugincenter.com). Source: EAA-PHEV.

CHAPTER 6

Engines and Electric Motors in the PHEV

If automobiles had improved as much as electronics in the past few decades, they would go a million miles per hour, cost only pennies, and last for decades.

Internal Combustion Engines

The term *internal combustion engine* (ICE) usually refers to an engine in which combustion is intermittent, such as the familiar four-stroke and two-stroke engines with pistons.

Fossil Fuels

Free-piston engines could be used to generate electricity as efficiently as, and less expensively than, fuel cells.[1]

Gasoline

Gasoline engines are used in most hybrid electric designs, and are likely to remain dominant for the foreseeable future. While petroleum-derived gasoline is the primary fuel, it is possible to mix in varying levels of ethanol created from renewable energy sources. Like most modern ICE-powered vehicles, HEVs can typically use up to about 15 percent bioethanol. Manufacturers may move to flexible-fuel engines, which would increase the allowable ratios, but no plans to do so are in place at present.

Diesel (and Biodiesel)

Diesel electric HEVs use a diesel engine for power generation. Diesels have advantages when it comes to delivering constant power for long periods of time, as they suffer less wear while operating at higher efficiency. The diesel engine's high torque, combined with hybrid technology, may offer substantially improved mileage. Most diesel vehicles can use 100 percent pure biofuels (biodiesel), so they can use petroleum for fuel but do not need it (although mixes of biofuel and petroleum are more common, and petroleum may be needed for lubrication). If diesel electric HEVs were in use, this benefit would probably also apply to them. Diesel electric hybrid drivetrains have begun to appear in commercial vehicles (particularly buses); as of 2007, no light-duty diesel-electric hybrid passenger cars are currently available, although prototypes exist.

Integrated Motor Assist on the Hybrid Electric Cars

Integrated Motor Assist (commonly abbreviated as *IMA*) is Honda's hybrid car technology, found in the Insight, the Civic Hybrid, and the Accord Hybrid. IMA uses a DC motor in the parallel hybrid system, in which the ICE and the electric motors work in parallel.[3]

Manual Integrated Motor Assist

There is a manual IMA (MIMA) control project that provides manual control over the IMA (for assist and regenerative braking) on the Honda Insight (see Figure 6-4). While this is not a PHEV project, it might make it possible to utilize additional

FIGURE 6-1 Mercury Mariner hybrid electric engine/traction motor.

Figure 6-2 Advanced energy PHEV internal combustion engine.

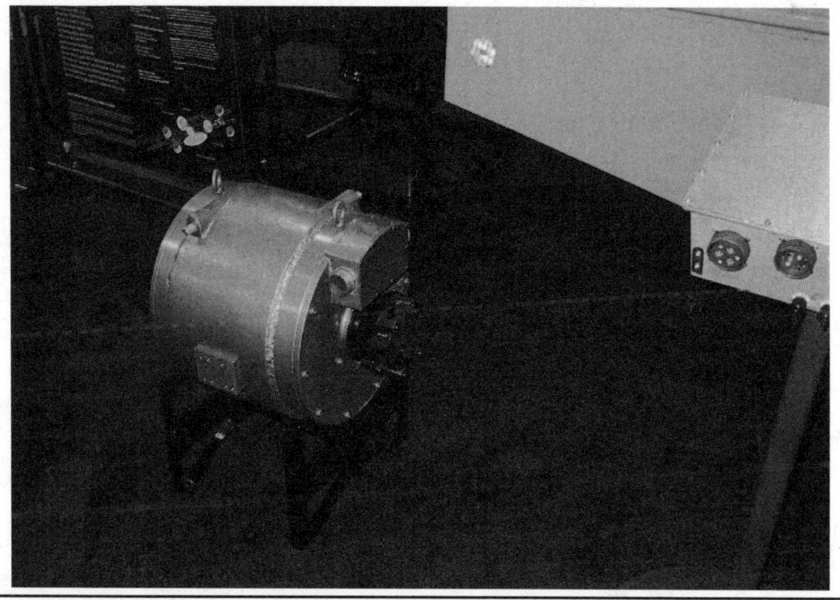

Figure 6-3 Advanced energy electric motor.

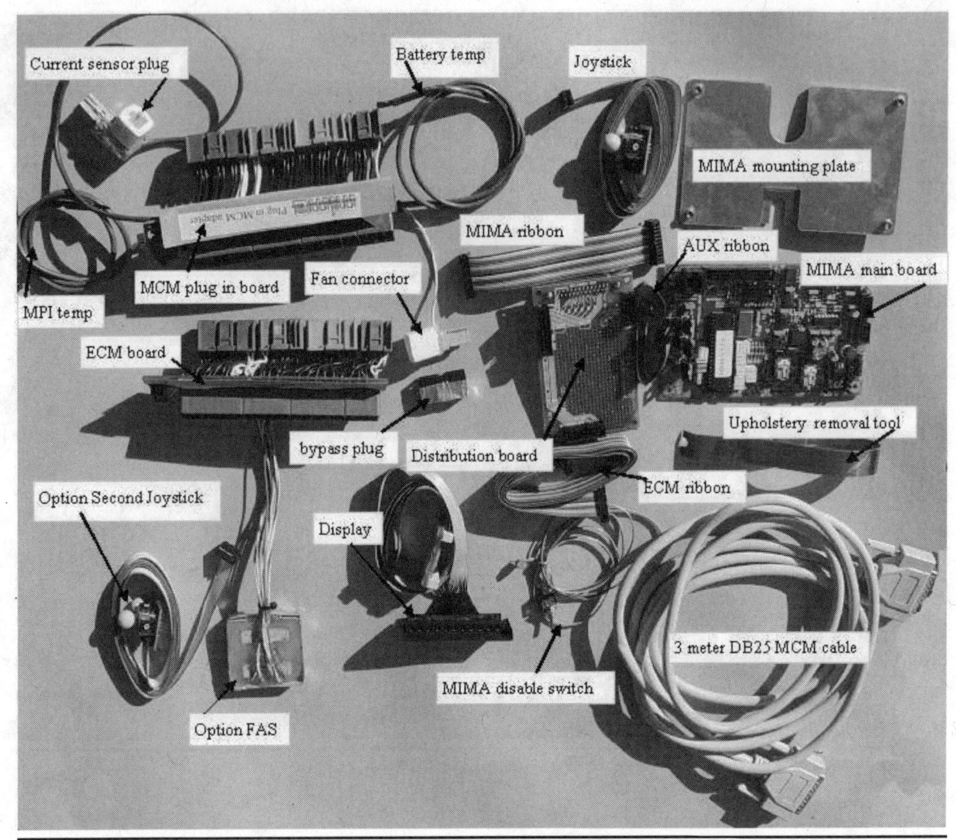

Figure 6-4 MIMA on the Honda Insight. Source: www.99mpg.com/mima/.

battery capacity in a mixed mode similar to that of Prius PHEVs at speeds greater than 34 mph.

MIMA would allow the driver to set an aggressive assist profile, allowing the IMA to contribute up to 10 kW of continuous power. This would rapidly deplete the stock battery, which might reach empty within five minutes or five miles. Some Insight drivers may already be familiar with recalibration events, or *recals*, which are essentially the manifestation of the state of charge (SOC) portion of the vehicle's battery management routines. While the most common recals are negative and occur when the SOC reaches a lower limit, it is also possible to recalibrate or drift up if the SOC is determined to be higher than expected. Such behavior makes it possible to manipulate the SOC simply by keeping the battery's voltage or perceived voltage near the upper limit.

Technically, as long as the lower SOC limit is not reached, MIMA makes it possible to command any desired level of IMA assistance, diminishing the importance of SOC manipulation. Both simple and hybrid battery pack configurations would be equally effective in such a conversion because of the previous point. A hybrid pack would leave the stock battery in place and might eliminate the need for a battery tap emulator, yet a simple configuration replacing the nickel–metal hydride pack with a superior lithium ion battery would maximize energy density and keep the vehicle's weight well under 2,000 pounds.[4]

Electric Motor

The heart of every electric vehicle and every PHEV is its electric motor. Unlike the gas engine, the electric motor emits no pollutants. In Figure 6-5, you can see that the electric motor for a PHEV is integrated directly into the wheel. This will give you the torque that you want out of your PHEV.

Voltage

Voltage is really called electromotive force.

To tie things into the electrical realm, Bob Brant showed us in *Build Your Own Electric Vehicle*, 2nd edition, that there is a mathematical equation that relates these parameters of force, flow, and resistance. The electrical equation, commonly known as Ohm's law, is

$$V = IR$$

where V is voltage in volts, I is current in amperes, and R is resistance in ohms. When you double the voltage, you send twice as much current through the wire, and the lightbulb becomes brighter. Or, if you reduce the resistance (as, to use a water analogy, can be done by enlarging a hole) while keeping the same voltage (the same level of water), you increase the current (increase the flow of water).

There are three moving parts in an electric motor, and electric motors usually outlive gas engines. The parts are the rotor and two end bearings.

The objective of this chapter is to guide you toward the best candidate motor for your PHEV conversion today, and to suggest the best type of electric motor for your future PHEV conversions.

What Is So Great about an Electric Motor?

The simplicity of electric motors is the secret of their dependability. In direct contrast to the internal combustion engine, with its hundreds of moving parts, electric motors are a far superior source of propulsion because they are inherently more powerful and more efficient than any other type of motor out there.

Horsepower

There is no way to explain horsepower other than telling about the first time I drove an electric car. When I stepped on that accelerator, it took off! No questions asked. There was no engine with excessive parts to get in the way of that.

Here are some technical points you need to understand when trying to find the right motor for your car.

DC Electric Motors

An electric motor is an ideal application of the fundamental properties of magnetism and electricity. Before looking at DC motors and their properties, let's review some fundamentals.

DC Motors in General

If you could support the conductor so that it could rotate in the magnetic field, you would create the condition shown in the upper part of Figure 6-5.

Now the current through the conductor exerts force that would tend to rotate it in the clockwise direction. The magnitude of the torque would be given by

$$T = Fr$$

where T is torque in foot-pounds, F is force in pounds, and r is the distance in feet, measured perpendicularly from the direction of F to the center of rotation. Now you have a motor design—on paper. You need to take some steps to make it real.

First, you need to make it more "force-full." Since the force varies with the length of the conductor, if you make a coil of wire, as shown in the upper right of Figure 6-5, twice as much length is cutting the lines of flux. The force generated on

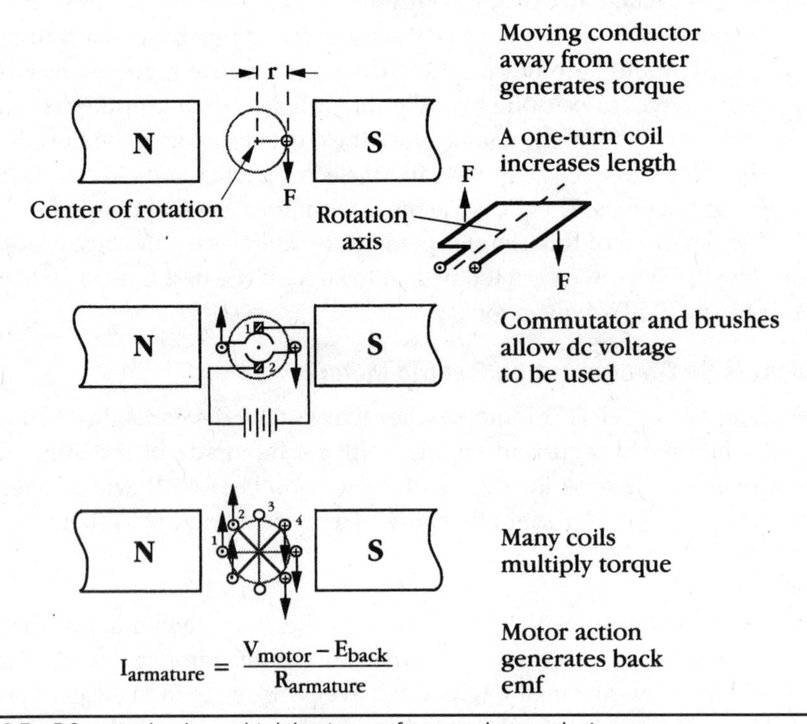

FIGURE 6-5 DC motor basics—obtaining torque from moving conductors.

the right-hand wire is downward, and the force generated on the left-hand wire is upward; they assist each other in producing rotation and result in twice the torque.

To further assist rotation, you add the commutator and brushes, shown in the middle of Figure 6-5. This arrangement allows you to power your motor from a constant supply of direct current (DC) voltage.

To further increase the motor's torque abilities, you can add additional coils, as shown at the bottom of Figure 6-5.

DC Motors in the Real PHEV World

Motor Case, Frame, or Yoke

In the motor, the magnetic path goes from the north pole through the air gap, the magnetic material of the armature, and the second air gap to the south pole, then back to the north pole again via the case, frame, or yoke.

Motors operating in the real world are subject to losses from three sources:

- *Mechanical.* Not all of the torque available inside the motor is available outside because torque is consumed in overcoming the friction of the bearings, moving air inside the motor, and brush drag.
- *Electrical.* Power is consumed as current flows through the combined resistance of the armature, the field windings, and the brushes.
- *Magnetic.* Additional losses are caused by eddy current and hysteresis losses in the armature and the field pole cores.

Types of DC Motors

Now that you've been introduced to DC motors in theory and in the real world, it's time to compare the different motor types. Figure 6-6 shows you a great way to view how the motors work. DC motors appear in the following forms:

- Series
- Shunt
- Compound
- Permanent-magnet
- Brushless
- Universal

Each of the motor types will be examined for its torque, speed, reversal, and regenerative braking capabilities—the factors that are important to EV users. The motor types will all be compared at full load shaft horsepower—the only way to compare different motor types of equal rating.

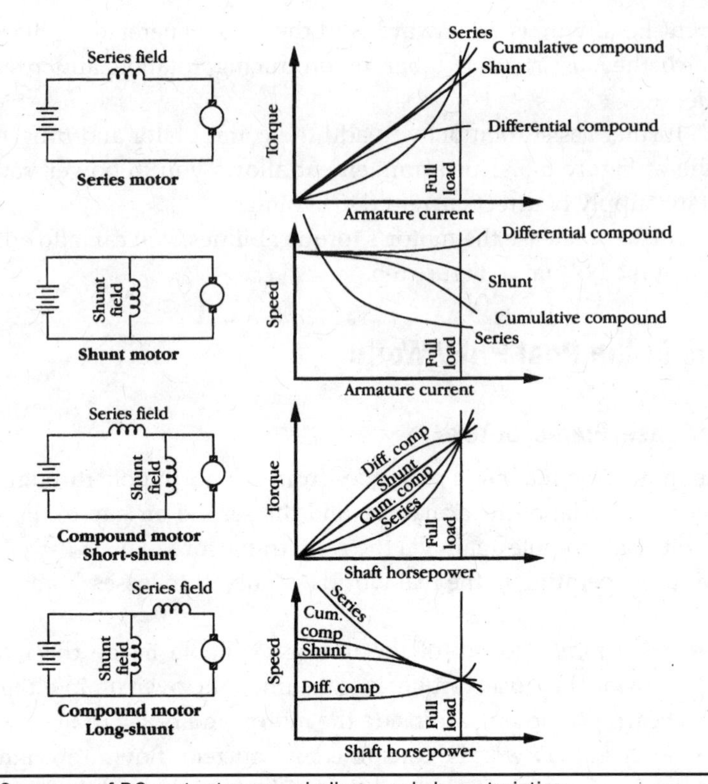

FIGURE 6-6 Summary of DC motor types—windings and characteristics.

Series DC Motors

The best-known type of DC motor, and the one that comes to mind for traction applications (like propelling EVs), is the series DC motor (Figure 6-6).

How the windings are made and connected determines the motor type. A coil of a few turns of heavy wire connected in series with the armature is called a *series motor*.

Torque

Torque = $K \times f \times I$, and the current in the series field is I. I can be substituted for f, so that the series motor torque equation can be written as

Torque = $K' \times I^2$

This shows that in a series motor, torque varies with the square of the current—a fact that is substantiated by the graph of actual torque versus armature current in Figure 6-6. In actual use, the armature reaction and magnetic saturation of the series motor at high currents set upper limits on both torque and current, although you might prefer to limit your circuit and components to far lower values. High starting torque makes series motors highly desirable for traction applications.

Speed

The equation for speed is

$$\text{Speed} = (V_t - I_a R_a)/K4)$$

Once again, V_t is the motor terminal voltage, I_a, R_a is the motor armature resistance and K4 is the constant incorporating all the motor fixed characteristics, can be substituted for *f*, and the series motor speed equation can be rewritten as

$$\text{Speed} = [V_t - I_a(R_a + R_f)]/KI_a$$

where R_a and R_f are the resistances of the armature and the field, respectively. This shows that in a series motor, speed becomes very large as the current becomes small. You need to make sure that you are always in gear, have the clutch in, have a load attached, and so on, because the series motor's tendency is to run away at no load. Just be aware of this and back off immediately if you hear a series motor rev up too fast.

Regenerative Braking

Regenerative braking allows you to slow down the speed of your PHEV (and save its brakes) and put energy back into its battery (thereby extending its drivable range) by harnessing its motor to work as a generator after it is up and running at speed.

Regenerative braking allows you to electronically switch the motor and turn it into a generator, thereby capturing the energy that would normally be dissipated (read: lost) as heat in the brake pads when you are slowing down. The motor does the braking, not your brake pads.

While all motors can be used as generators, the series motor is rarely used as a generator in practice because of its unique and relatively unstable generator properties.

Shunt DC Motors

A coil of many turns of fine wire connected in parallel with the armature is called a *shunt motor* (Figure 6-6). Because it doesn't have to handle the high motor armature currents, a shunt motor field coil is typically wound with many turns of fine-gauge wire and has a much higher resistance than the armature. These motors are great for regenerative braking, so don't think that DC motors are cheaper and can't do regen. It's just not the case!

Compound DC Motors

A compound DC motor is a combination of the series and shunt DC motors. The way its windings are connected and whether they are connected to boost (assist) or buck (oppose) one another in action determine its type. Its basic character is determined by whether current flowing into the motor first encounters a series field coil–short-shunt compound motor or a parallel shunt field coil–long-shunt compound motor, as shown in Figure 6-6. If, in either one of these configurations, the coil windings are hooked up to oppose one another in action, you have a differential compound motor. If the coil windings are hooked up to assist one another in action, you have a cumulative compound motor. The beauty of the compound motor is its ability to bring the best of both the series and the shunt DC motors to the user. A compound motor is as easily adaptable to regenerative braking as a shunt motor. Its series winding gives it additional starting torque, but this can be bypassed during regenerative braking applications.

The torque in a compound motor has to reflect the actions of both the series and the shunt field coils. The effect of these hookup arrangements on torque is illustrated in Figure 6-6, where the differential compound motor builds more slowly to a lower torque value than the shunt curve and the cumulative compound motor builds to a slightly higher torque value than the shunt curve at a slightly higher rate.

Figure 6-6 also shows the speed curves. One of the initial benefits of the compound configuration is that it can eliminate runaway conditions at low field current levels for the shunt motor and at lightly loaded levels for the series motor. You can tailor a cumulative compound motor to your PHEV needs by picking one whose series winding delivers good starting torque and whose shunt winding delivers lower current draw and regenerative braking capabilities once it is up to speed. When you look, you might find that these characteristics already exist in an off-the-shelf model.

Permanent-Magnet DC Motors

Permanent-magnet motors are being increasingly used today because new technology—various alloys of Alnico magnet material, ferrite-ceramic magnets, and rare-earth-element magnets—enables them to be made smaller and lighter in weight than equivalent wound field coil motors with the same horsepower rating. Rare-earth-element magnets surpass the strength of Alnico magnets significantly (by 10 to 20 times) and have been used with great success in other areas, such as computer disk drives, thereby helping to drive down the production costs. (See Figure 6-7.)

Permanent-magnet motors approximately resemble the shunt motor in their torque, speed, reversing, and regenerative braking characteristics; either motor type can usually be substituted for the other in control circuit designs. Permanent-magnet motors have starting torques several times those of shunt motors, and their speed-versus-load characteristics are more linear and easier to predict.

Brushless DC Motors

Brushless motors promise to be the most long-lived and maintenance-free of all motors. Assume that brushless DC motors resemble their permanent-magnet DC motor cousins in characteristics—those of a shunt motor plus high starting torque

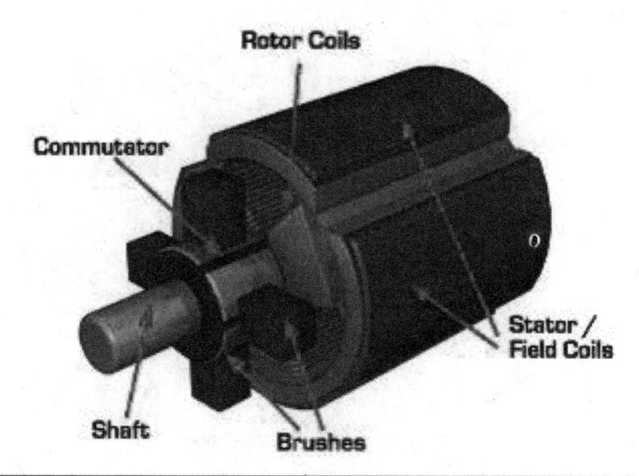

Figure 6-7 Permanent-magnet DC motor. Courtesy of Zero Emission Vehicles of Australia and www.electric-cars-are-for-girls.com.

plus linear speed/torque—with the added kicker of even higher efficiency because there are no brushes (see Figure 6-8).

Universal DC Motors

DC motors designed to run on AC typically have improved lamination field and armature cores to minimize hysteresis and current losses (see Figure 6-9). Series DC motors operating on AC perform almost the same (high starting torque, and so on), but are less efficient.

AC Electric Motors

Homes, offices, and factories are fed by alternating current (AC). Since it can easily be transformed from high voltage for transmission into low voltage for use, more AC motors are in use than all the other motor types put together. Before looking at AC motors, however, you need to look at transformers.

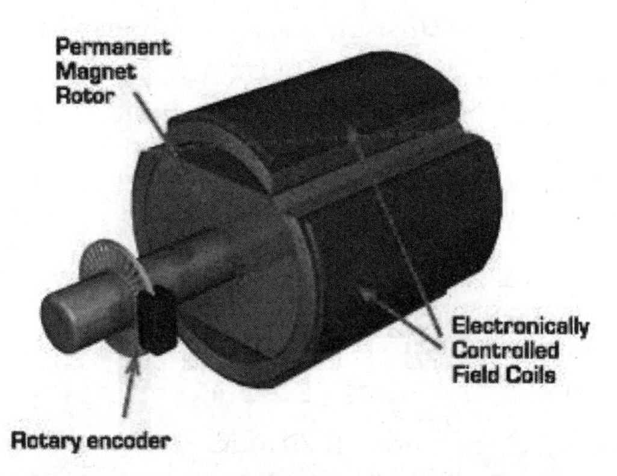

Figure 6-8 Brushless DC motor. Courtesy of Zero Emission Vehicles of Australia and www.electric-cars-are-for-girls.com.

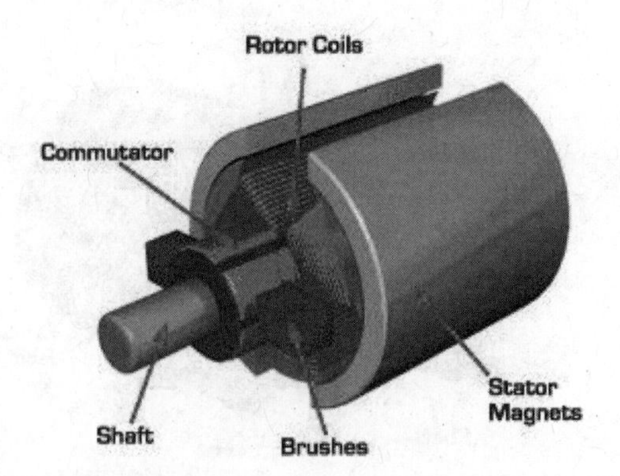

Figure 6-9 DC motor with round stator. Courtesy of Zero Emission Vehicles of Australia and www.electric-cars-are-for-girls.com.

Transformers

In its simplest and most familiar form, a transformer consists of two copper coils wound on a ferromagnetic core (Figure 6-10). The primary is normally connected to a source of alternating electric current. The secondary is normally connected to the load. The other aspect of transformers that is useful to you is that an equivalent circuit of a transformer can be drawn for any frequency, and you can study what is going on. This is useful and directly applicable to AC induction motors.

AC Induction Motors

The AC induction motor, patented by Nikola Tesla back in 1888, is basically a rotating transformer. While AC motors come in all shapes and varieties, the AC induction motor—the most widely used variety—holds the greatest promise for EV owners because of its significant advantages over DC motors.

The most common split-phase induction motor is one that uses a capacitor start, also shown in Figure 6-10. The capacitor automatically provides a greater electric phase difference than inductive windings. The principle was discovered by Charles Steinmetz and others, but capacitor technology had to catch up before it could be widely introduced on production motors. Capacitor-start design variations include two types: separate starting and running capacitors, and a permanent capacitor with no centrifugal cutout switch. The two-capacitor approach brings you the best of both the starting and the running worlds; the permanent capacitor type gives you superior speed control during operation at the expense of lower starting torque.

Polyphase AC Induction Motors

Polyphase means having more than one phase. AC is the prevailing mode of electricity distribution. The phase voltage that comes from the pole is 240 V. This voltage is widely available in nearly every city in the industrialized world.

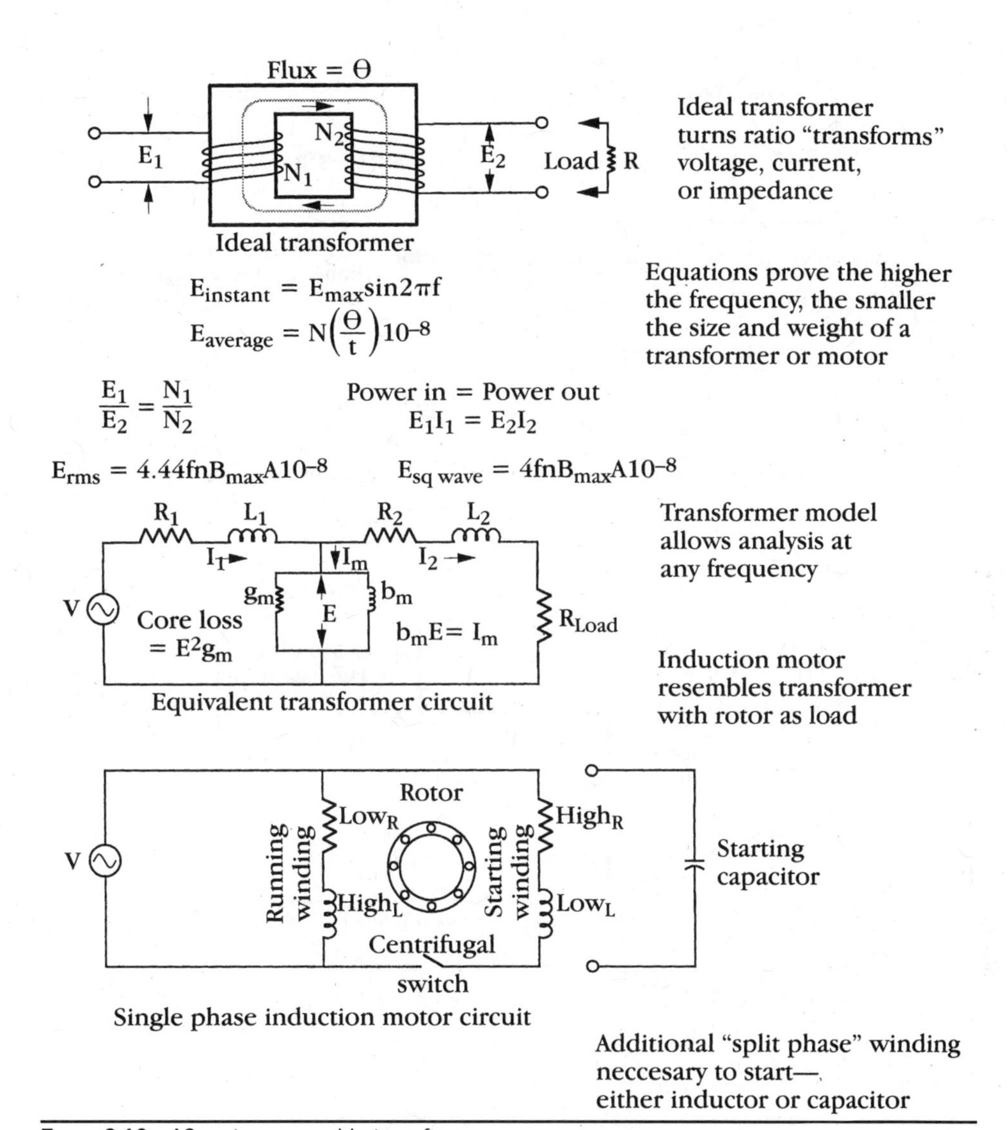

FIGURE 6-10 AC motors resemble transformers.

Stationary three-phase electric induction motors are inherently self-starting and highly efficient, and electricity is conveniently available.

Three-phase AC connected to the stator windings of a three-phase AC induction motor produces currents that look like those shown at the top of Figure 6-11—they are of the same amplitude, but 120 degrees out of phase with one another. As in a DC motor, power and torque are also a function of current in an induction motor.

The characteristic induction motor torque-to-slip graph, shown in Figure 6-11 for both its motor and its generator operating regions, offers insight into induction motor operation. Also, speed and torque are relatively easy to handle and determine in an induction motor. So are reversing and regenerative braking.

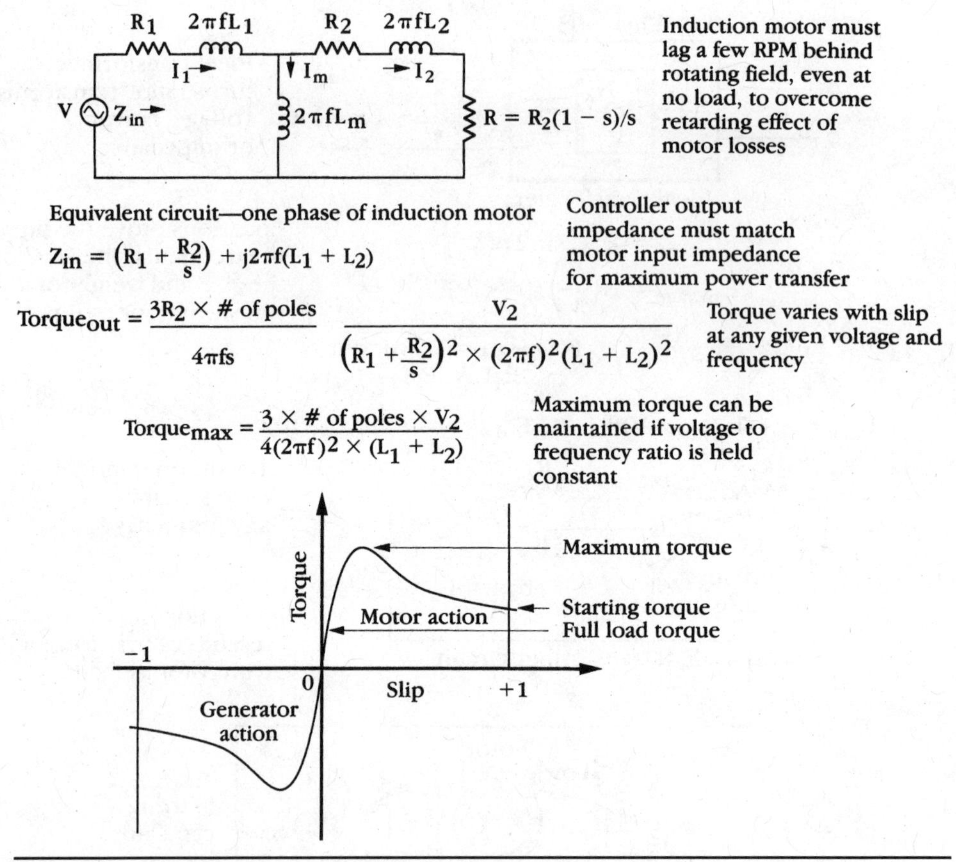

$$Z_{in} = \left(R_1 + \frac{R_2}{s}\right) + j2\pi f(L_1 + L_2)$$

$$Torque_{out} = \frac{3R_2 \times \#\text{ of poles}}{4\pi fs} \quad \frac{V^2}{\left(R_1 + \frac{R_2}{s}\right)^2 \times (2\pi f)^2(L_1 + L_2)^2}$$

$$Torque_{max} = \frac{3 \times \#\text{ of poles} \times V^2}{4(2\pi f)^2 \times (L_1 + L_2)}$$

FIGURE 6-11 Polyphase AC motor's unique speed, torque, and slip characteristics versus voltage and frequency.

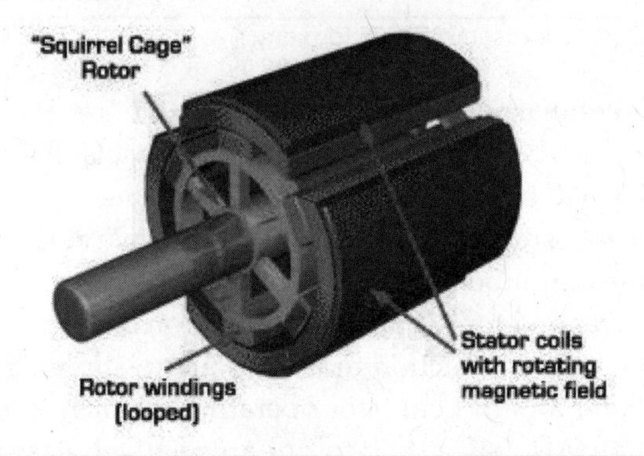

FIGURE 6-12 AC electric motor. Courtesy of Zero Emission Vehicles of Australia.

Wound-Rotor Induction Motors

A wound rotor's windings are brought out through slip rings—conductive rings on the rotor's shaft—and through brushes (analogous to DC motor construction) to an external resistance in series with each winding. The advantage of the wound-rotor induction motor over the squirrel-cage induction motor is that resistance control can be used to vary both the motor's speed and its torque characteristics. Increasing the resistance causes maximum torque to be developed at successively higher values. What you give up is efficiency, and complexity and cost are increased.

Today's Best PHEV Motor Solution

For economy conversions, the DC series-wound motor has been found to be the best approach. Other motors are getting harder to find, and not every controller on the market can drive them. For efficiency and for regenerative braking (which buys about 25 percent additional range in city driving), AC motors are the best way to go.

As I stated in *Build Your Own Electric Vehicle*, series DC motors are available from many sources, they work well, controllers are readily available, adapters to different vehicles are easily made or purchased for them, and the price is right. A series DC motor might not be the ultimate or even the best current solution, but it's one that most PHEV converters will have no trouble implementing today.

Conclusion

Currently, the most economical (and indeed the most common) option for plug-in hybrid electric vehicles is series DC technology. At present, AC induction and permanent-magnet brushless DC are the best technologies available, with efficiencies of up to 98 percent, silent operation, and almost never any need for servicing. Each has various advantages and disadvantages compared to the other. It will be interesting to see which one becomes the new standard in the years to come.

CHAPTER 7

The Controller

"The superior AC system will replace the entrenched but inferior DC one."
—George Westinghouse *(from Tesla: Man Out of Time)*

The controller is another pillar of every plug-in hybrid electric vehicle. If one element were to take credit for renewed interest in PHEVs, the controller would be a prime candidate.

You can buy

- A controller that runs the entire plug-in hybrid system
- A separate controller that communicates with the main controller (for the hybrid car) and also communicates with the extra battery packs in the system

In this chapter, you'll learn what the different types of controllers are, how they work, and their advantages and disadvantages. Then you'll discover the best type of controller to choose for your PHEV conversion today (the type used in Chapter 10's conversion) and the electric motor controller that you're likely to be seeing a lot more of in the future.

Controller Overview

Basically, the controller is the brain or computer of an electric car. This computer "controls" or governs the performance of the electric motor. It communicates with the electric motor and battery packs (through the energy density), allowing you to go from 0 to 60 in 6 to 7 seconds (or less). A controller that is correctly matched to the motor will give it the right voltage. The weight of the motor magnets and the size of the motor's brushes determine the power and torque.

Solid-State Controllers

The creation and wide availability of good solid-state controllers has made the PHEV market practical and everyday rather than something that will develop in the future.

In *Build Your Own Electric Vehicle*, I said that you can regard a good controller as a black box, without having to understand the insides to the *n*th detail. This means (unfortunately) that detailed technical knowledge is not needed. On the other hand, for some people, it's fortunate that technical knowledge is not needed. Figure 7-1 shows a typical PHEV battery pack controller, and Figure 7-2 shows the insides of that controller.

OK, higher-end components with an AC controller from AC Propulsion can cost $25,000. However, most parts for the PHEV cost thousands of dollars.

Digital control can outdo what analog electronics did by leaps and bounds. Automakers resorted to fuel injection rather than continuing with carburetors because of the much more precise control that fuel injection allowed and in order to meet emissions requirements. Digital control of motor speed can work on a per piston firing basis in a hybrid. The IMA (Integrated Motor Assist) on the Honda hybrid electric cars was used to torque the electric motor to smooth out rough engine idling on the 2000 Insight (no longer in production), and the user never understood that the three-cylinder engine was idling poorly.

Low maintenance and hassle-free usage at minimum cost are the primary reasons to purchase the best controllers available. There is no way under the sun that a hobbyist can duplicate the reliability of the commercially made hardware.

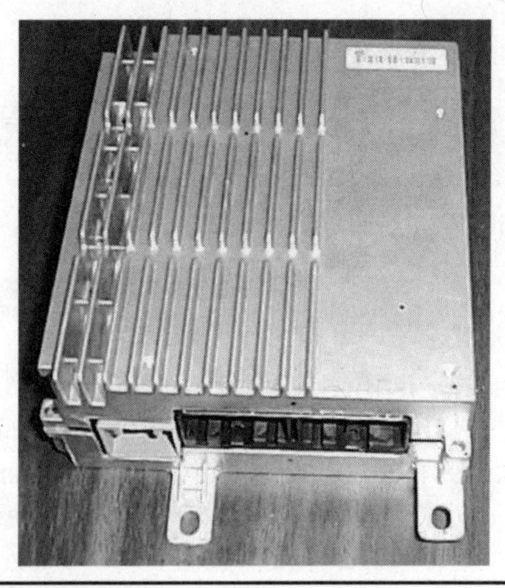

Figure 7-1 Typical PHEV battery controller.

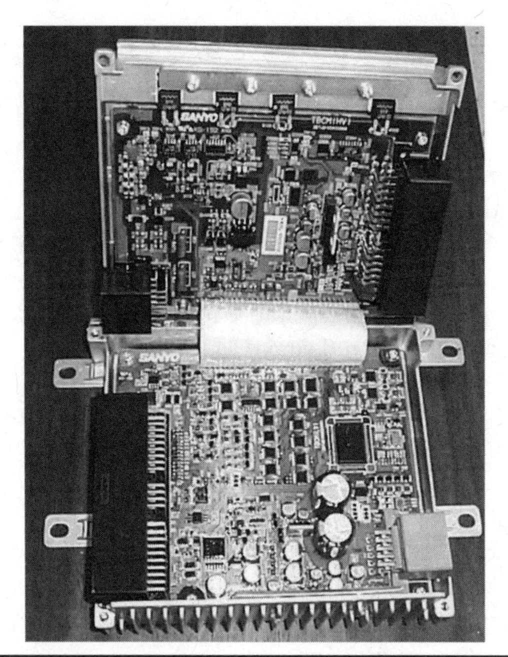

FIGURE 7-2 Inside the PHEV battery controller.

Imagine having to tinker with the electronics every morning before you headed out for the day. Unacceptable!

Armed with this component information, you are now ready to look at the motor controllers that are available. Let's start with the simplest first.

Battery Management Systems

A battery management system (BMS)[1] is used to manage a battery pack, such as by

- Monitoring its state
- Calculating secondary data
- Reporting those data and protecting the battery
- Controlling its environment
- Balancing it

A BMS may monitor the state of the battery by measuring various items, such as

- *Voltage:* total voltage, voltage of periodic taps, or voltages of individual cells
- *Current:* current into or out of the battery
- *Temperature:* overall pack temperature, air intake or exhaust temperature, or individual cell temperatures
- Environmental conditions, e.g., air flow in air-cooled batteries

Additionally, a BMS may calculate values based on these items, such as

- State of charge (SOC) or depth of discharge (DOD) to indicate the charge level of the battery
- State of health (SOH), a variously defined measurement of the overall condition of the battery
- Maximum charge current as a charge current limit (CCL)
- Maximum discharge current as a discharge current limit (DCL)
- Resistance: dynamic resistance for the entire pack or for individual cells
- Total energy delivered since manufacture
- Total operating time since manufacture

A BMS may report all of these data to an external device, using communication links such as

- A CAN bus (typical of automotive environments)
- Direct wiring
- Serial communications
- Wireless communications

A BMS may protect the battery by preventing it from operating outside its safe operating range, avoiding such problems as

- Overcurrent
- Overvoltage (during charging)
- Undervoltage (during discharging); this is especially important for lead-acid and lithium ion cells
- Overtemperature or undertemperature
- Overpressure (typical of nickel–metal hydride batteries)

The BMS may prevent the battery from operating outside of its safe operating area by

- Requesting that the devices to which the battery is connected reduce or even terminate their use of the battery
- Including an internal switch (such as a relay or solid-state device) that is opened if the battery is operated outside of its safe operating area
- Actively controlling the environment, such as through heaters, fans, or even air conditioning

In order to maximize the battery's capacity and to prevent localized undercharging or overcharging, the BMS may actively ensure that all the cells that make up the battery are kept at the same state of charge. It may do so by

- Wasting energy from the cells with the highest charge through a dummy load (regulators)
- Shuffling energy from the cells with the highest charge to those with the lowest (balancers)
- Reducing the charging current to a sufficiently low level that it will not damage fully charged cells, while allowing less charged cells to continue to charge

BMS Technology Complexity and Performance Range

- Simple passive regulators across cells bypass the charging current when their cell's voltage reaches a certain level to achieve balancing.
- Active regulators intelligently turn on a load when appropriate, again to achieve balancing.
- A full BMS reports the state of the battery to a display and protects the battery.

BMS topologies mostly fall into three categories:

- *Centralized:* a single controller is connected to the battery cells by a multitude of wires.
- *Distributed:* a cell board is installed at each cell, with just a single communication cable between the battery and a controller.
- *Modular:* a few controllers, each handling a certain number of cells, communicate with one another.

Centralized BMSs are most economical, least expandable, and plagued by a multitude of wires (spaghetti). Distributed BMSs are the most expensive and the simplest to install, and offer the cleanest assembly. Modular BMSs offer a compromise between the features and problems of the other two topologies.[2]

The company Elithion has developed a great BMS platform for managing the battery pack.

AC Controllers

Alternating current has benefits that make it overwhelmingly a winner, in spite of the complications involved. In general, AC motor controllers require more protection devices to isolate against noise, yet DC motors make far more noise than AC!

Chapter 6 showed that the speed-torque relationship of a three-phase AC induction motor is governed by the amplitude and frequency of the voltage applied to its stator windings (the upper left part of Figure 7-4 depicts this relationship). The best way to change the speed of an AC induction motor is to change the frequency of its stator voltage. As you can see in Figure 7-4, a change in frequency results in a direct change in speed, and if you change the frequency in proportion to the voltage (both at ¼, ½, ¾, etc.), you get the speed-torque curves shown.

Elithion Controller Package[3]

Figure 7-3 shows a battery management system (BMS) controller at the board level. Let's see some guts, why don't we? Don't worry, Figure 7-4 shows the controller closed to show the casing. Figure 7-5 shows pretty much the same thing as Figure 7-3, but without the notes next to the parts. This will help train you to learn the BMS without the notes. If you don't care to do that, Figure 7-4 is your friend because it's just plug and play!

Selector

Use the flowchart in Figure 7-6 to select a combination of components for the BMS controller set, based on your system needs.

The BMS controller must read the battery current using one of these methods:

- The battery current value is already present on the CAN bus; neither a high-voltage front end nor a cable-mounted current sensor is required.
- A high-voltage front end, which includes a current sensor, is connected to the BMS controller.
- A cable-mounted current sensor is connected to the BMS controller.

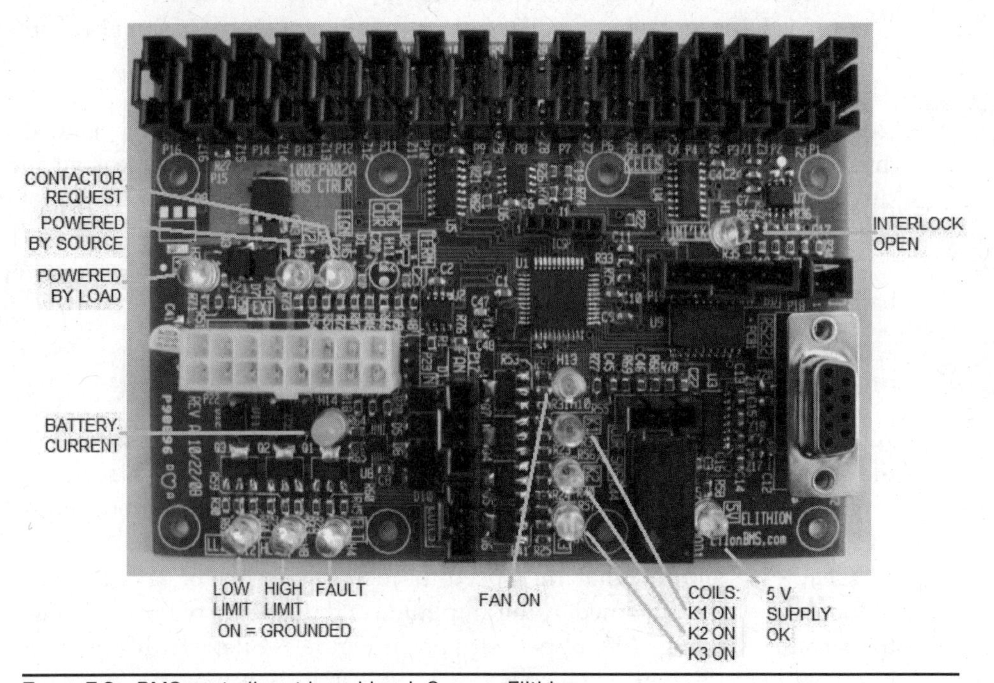

FIGURE 7-3 BMS controller at board level. Source: Elithion.

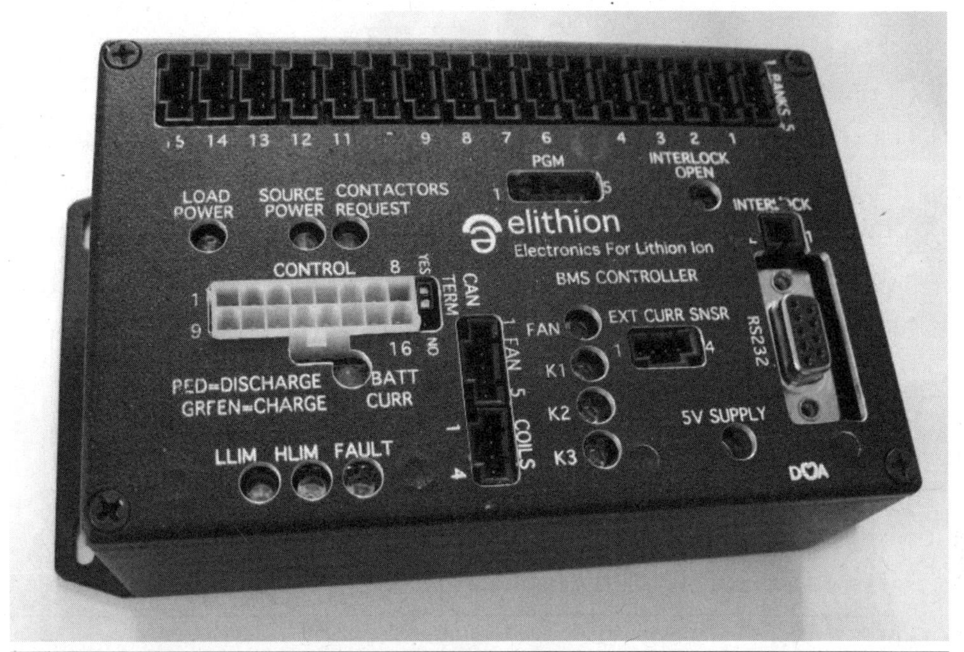

Figure 7-4 Enclosed Elithion BMS controller. Source: Elithion.

Figure 7-5 BMS controller board without notes like those in Figure 7-3.

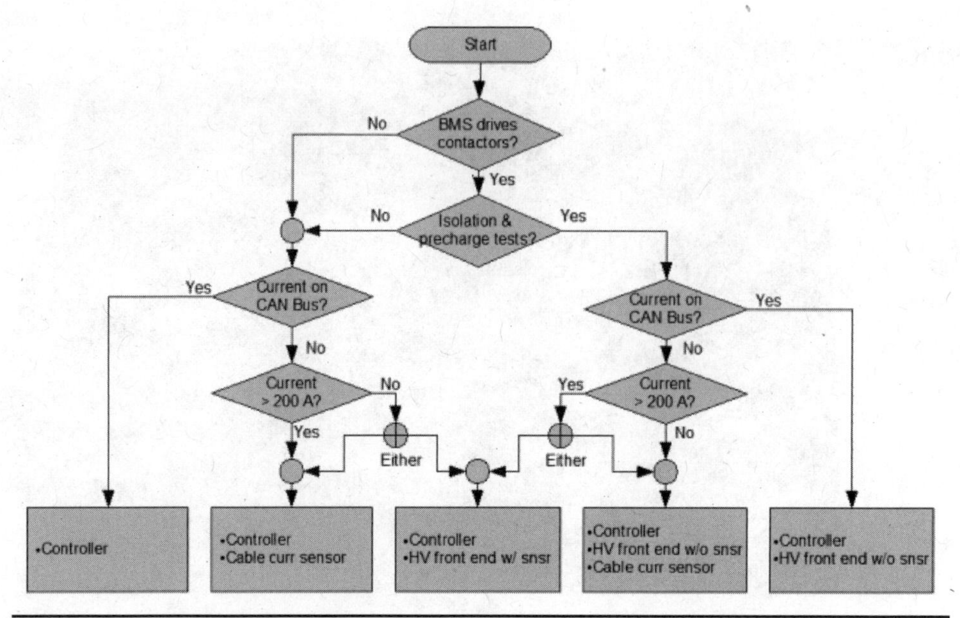

FIGURE 7-6 Flowchart for the BMS controller.

This is not required if the BMS controller includes a high-voltage front end or if there's no need to check the end of precharge and loss of isolation.

A cable-mounted current sensor is slipped onto a high-voltage battery cable and connected to a BMS controller.

If a cable-mounted current sensor is used, then either there must be no high-voltage front end or the high-voltage front end must not include a current sensor. This optional current sensor, when combined with a BMS controller without current sensors, completes a lithium ion BMS.

Separate High-Voltage Front End for Controller

The PHEV component for the optional and separate high-voltage front end includes circuits that are connected to the battery and are therefore at high voltage. This optional high voltage front end, when combined with a BMS controller, completes a lithium ion BMS by adding safety testing and may include current sensing. Figure 7-7 shows the high-voltage circuits that are added to the controller. Putting those circuits on a separate board improves safety in the BMS controller and reduces the chance of electrical noise interference with it. This board includes the following functions:

- Detector of loss of isolation to ground
- Detector of current through the precharge resistor
- Optionally, an onboard sensor for battery current
- Optionally, a cable-mounted current sensor

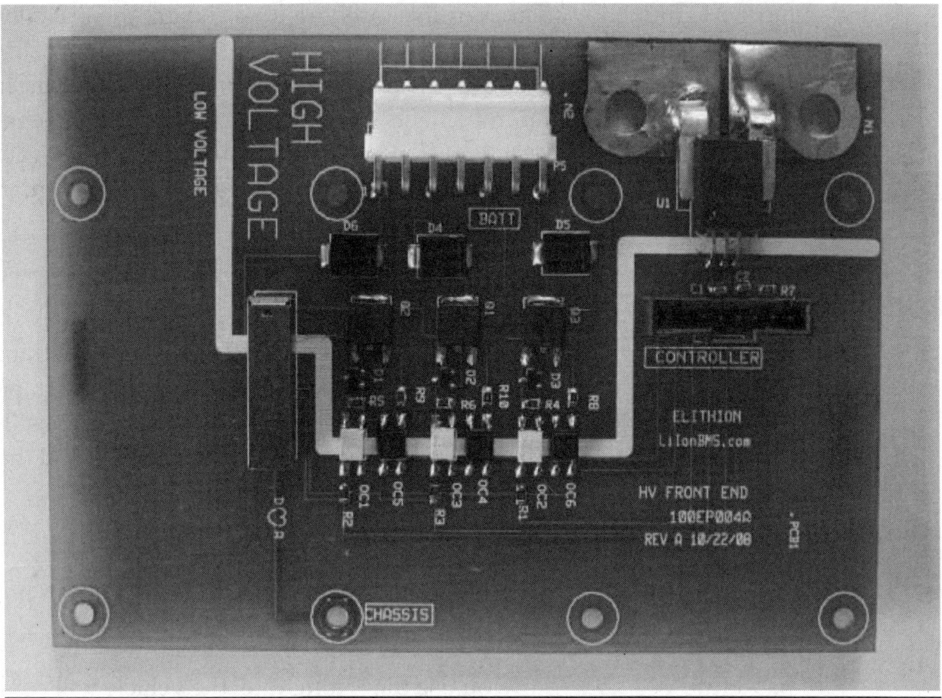

Figure 7-7 Controller board level hybrid vehicle front end. Source: Elithion.

A separate high-voltage front end

- Keeps the high voltage away from the BMS controller
- Keeps any electrical noise away from the BMS controller
- May simplify the layout of high-voltage cabling

Separate high-voltage front ends are available

- At board level or encased in a metal enclosure
- With or without internal current sensors (of various ranges)[4]

DC Controllers

While the series DC motor and pulse width modulation controller are unquestionably best for today's first-time PHEV converters, the bias of this book is toward AC controllers and motors. AC induction motors are inherently more efficient, more rugged, and less expensive than their DC counterparts. This translates into more driving range from a given set of batteries, less probable failures, and the possibility of graceful degradation when a failure does happen. The package isn't cheap, and the motor alone isn't either. These benefits come at a price. That's why nearly every newly designed commercial PHEV today utilizes one or more AC induction motors or its closely related cousin, the brushless DC motor.

What's in the labs today will be available to you in the not-too-distant future, and, beyond that, continued improvements in solid-state AC controller technology could put AC motors in every PHEV conversion of the future. Let's look at developments in two areas—systems and components—that virtually guarantee this outcome. Now, let's look at the future—today!

2CH00xxx BMS Controllers CAN Specs

These are the specifications for the messages on the 2CH00xxx standard lithium ion BMS controllers and the controller area network (CAN) bus.[5] This is an important section because it defines a lot of aspects of the controllers and the CAN bus. That is why this section lists and describes the CAN messages sent by the BMS and the ones that the BMS listens to.

CAN Bus Protocol

The CAN bus is active only when the vehicle is in on (in the Start or Run position of the ignition key) and for a few seconds after the vehicle is turned off.

- Baud rate: 500 kbits/s (if you use the wrong rate, the vehicle will complain and store a DTC fault until the DTC codes are cleared)
- Standard: CAN 2.0A ("standard CAN," 11-bit identifier)
- Remote frames: not used

This means that all the data are volunteered and none are requested; that is, every component on the vehicle broadcasts its data periodically—no component puts out requests for data.

CAN Tools[6]

A generic adapter is needed between the CAN bus and a PC. It is convenient to use a USB port, although the serial, parallel, or Ethernet port may be used as well.

Note: Disconnecting a product's USB cable seems to create significant problems for Windows XP (immediate shutdown, or even the "blue screen of death"). You must use the system tray's "Remove hardware" icon first.[7]

The Prius's OBD (On Board Diagnostics) connector (also known as the Data Link Connector 3, or DLC3) is located under the dashboard, below and to the right of the steering wheel, facing down.

Alternatively, you can tap into the CAN bus directly. Use a short cable to the CAN adapter.

- CAN –: black wire
- CAN +: white wire
- GND: chassis

Transmitted CAN Messages

The BMS controller places the following messages[8] on the CAN bus:

- Standard traction pack messages
- Control message for a BRUSA charger
- Parameter messages for RM Michaelides

CAN Displays

The BMS controller operates on these fixed parameters:

- CAN communication standard: 500 kHz
- Standard addressing (CAN 2.0A) (not extended)

Standard Traction Pack Messages

These are the messages generated by the battery.

The battery ECU (electronic control unit) broadcasts the messages given in Table 7-1. In this table, numbers in parentheses (#) refer to the notes just below the table. Names in parentheses are hunches.[9]

The Elithion BMS controller places the applicable subset of the standard traction pack messages on the CAN bus.

Notes:

- Period: 1 s
- Multibyte values are big-endian: the MSB (most significant byte) is in the lower-numbered data byte (the leftmost byte in Tables 7-1 to 7-3).
- The ID of the first message is programmable; the other messages use IDs following the first ID.

Battery Current Messages from Elithion

The BMS controller looks on the CAN bus for messages with the battery current from/to the load and the source.

The format is programmable.

Notes:

- Maximum period: 300 ms (after which it assumes that there is no such message on the CAN bus)
- Default messages

State of PHEV System

- 2: Powered from the load, unable to discharge
- 4: Powered from the load, able to discharge
- 8: Powered from the source, unable to charge

TABLE 7-1 Standard Traction Pack Messages

ID	Bytes	Byte 0	Byte 1	Byte 2	Byte 3	Byte 4	Byte 5	Byte 6	Byte 7
620h	8	Company name							
621h	8	Product name/rev level							
622h	6	State	Timer		Flags	DTC	DTC	—	
623h	6	Voltage		Min vtg	Min vtg #	Max vtg	Max vtg #	—	
624h	6	Current		Charge limit		Discharge limit		—	
625h	8	Battery energy in				Battery energy out			
626h	6	SOC	DOD		Capacity		—	SOH	—
627h	6	Temperature	—	Min tmp	Min tmp #	Max tmp	Max tmp #	—	
628h	6	Resistance		Min res	Min res #	Max res	Max res #	—	

Eight ASCII characters: (e.g.:"Elithion", "2CH00xxx")

- 10: Powered from the source, able to charge
- 15: Powered by both source and load, contactor request line input is high: driving off while plugged in
- Count-up timer: time in the present state. 255 s max.

Byte of Flags

- Bit 0: Receiving power from the load
- Bit 1: Receiving power from the source
- Bit 2: Contactor request line input is high
- Bit 3:
- Bit 4: Fault
- Bit 5: High limit: battery is full
- Bit 6: Low limit: battery is empty
- Bit 7: Fan on

Error Codes: Reserved

- Total voltage of pack (V), unsigned, 0 to 65 kV.
- Voltages (100 mV) of least-charged and most-charged cells or blocks of cells, 0 to 25.5 V.
- ID of the cell (or block of cells) that has the lowest/highest voltage/temperature/resistance, 1 to 254.
- Pack current (A), signed, positive out of pack, –32 kA to +32 kA.
- Maximum current acceptable (charge) or available (discharge) (A), unsigned, 0 to +65 kA.
- Total energy into or out of battery since manufacture (Wh), unsigned; overflows back to 0.
- State of charge (%), unsigned, 0 to 100. When deeply discharged, its value does not go below 0.
- Depth of discharge (Ah), unsigned, 0 to 65 kAh. When deeply discharged, its value may exceed the actual capacity value.
- Actual capacity of pack (Ah), unsigned, 0 to 65 kAh.
- Not used.
- State of health (%), unsigned, 0 to 100; 100% = all OK.
- Average pack temperature (°C), signed, –127°C to +127°C.
- Temperatures (°C) of coldest and hottest sensors, signed, –127°C to +127°C.
- Resistance of pack (mohm), unsigned, 0 to 65 ohms.
- Resistances (100 micro-ohms) of lowest- and highest-resistance cells (or blocks of cells), unsigned, 0 to 25.5 milliohms.

BRUSA Charger Control Message

This BMS controller places a control message for an NLG5 charger on the CAN bus.

Notes:

- Period: 100 ms
- Multibyte values are big-endian: the MSB (most significant byte) is in the lower-numbered data byte (the leftmost byte in the tables)
- Control flags: only 1 bit is used; all others are 0:
  - Bit 7: 1 to enable charging; 0 to disable charging
- Maximum current from the AC inlet (100 mA)
- Maximum DC output voltage (100 mV)
- Maximum DC output current (100 mA)

Table 7-2 Control Messages for the Battery Charger

ID	Bytes	Byte 0	Byte 1	Byte 2	Byte 3	Byte 4	Byte 5	Byte 6	Byte 7
618h	7	Flags	Max I_{AC}		Max V_{DC}		Max I_{DC}		

Table 7-3 CAN Bus Messages; Source: Elithion.

ID	Bytes	Byte 0	Byte 1	Byte 2	Byte 3	Byte 4	Byte 5	Byte 6	Byte 7
680h	2	SOC Bar	Status LEDs	—					

Display messages on the CAN bus.

The CAN bus has only 24 messages.

The screen capture in Table 7-4 is from a 2007 Hybrids Plus Escape. The 2008 Hybrid Escape has more messages: 41h, 350h.

Battery Management System (BMS) Controller

This BMS controller places a message for an Elithion 6DS000xK display on the CAN bus.

Notes:

- Period: 100 ms
- State of charge (%)
- Status LEDs, 1 = on
  - Bit 0: bottom LED, blue
  - Bit 1: second LED, green
  - Bit 3: third LED, yellow
  - Bit 4: fourth LED, amber
  - Bit 5: top LED, red

TABLE 7-4 Screen Capture Taken with the Ignition On, Engine Off, from a 2007 Hybrids Plus Escape; Source: EAA-PHEV.

Me...	Length	Data	Period
040h	6	00 00 00 00 00 00	100
046h	6	00 00 00 00 00 00	8
170h	8	0F A0 6E 3C 00 00 14 00	8
180h	3	00 00 00	16
300h	5	05 DC 93 00 00	10
310h	7	8C 78 50 3C 68 71 86	100
312h	8	7F FF 00 7F FF 03 7F FF	10
320h	5	00 00 01 01 F2	100
325h	2	64 00	20
326h	8	00 80 00 00 00 00 00 00	15
332h	8	7F FF 7F FF 7F FE 00 C0	8
333h	1	00	50
400h	7	20 00 3C 00 00 A0 07	10
420h	8	00 00 0A 00 1F 40 04 4C	16
422h	2	00 1E	48
425h	8	00 00 23 28 02 58 00 00	16
430h	8	FC E7 00 00 00 00 00 00	100
450h	7	1F 40 00 00 0F A0 00	16
470h	7	37 37 37 40 3F 00 00	100
4B0h	8	27 10 27 10 27 10 27 10	10
4FFh	8	30 31 46 4D 59 55 00 79	248
500h	8	46 14 39 37 64 00 00 53	96
510h	3	46 10 24	48
575h	8	00 00 00 00 00 00 00 00	16

Received CAN Messages

TABLE 7-5 Default Messages from Elithion

ID	Bytes	Byte 0	Byte 1	Byte 2	Byte 3	Byte 4	Byte 5	Byte 6	Byte 7
611h	8							Source current	
633h	8	Load current							

Notes:

- (10 mA), + = into battery, big-endian. Preset for compatibility with BRUSA NLG5 charger.
- (100 mA), + = out of battery, big-endian.

Contactor Control Message

This BMS controller looks on the CAN bus for messages to control the contactors. The format is programmable.

Notes:

- Maximum period: 300 ms (after which it assumes that there is no such message on the CAN bus)

TABLE 7-6 Default Messages

ID	Bytes	Byte 0	Byte 1	Byte 2	Byte 3	Byte 4	Byte 5	Byte 6	Byte 7
632h	8	Bit 0: Contactor request (1)							

- 1: requests contactors be on; 0: requests contactors be off

Request and Response CAN Messages—PID Support

- Request is at ID 07E3h.
- Data bytes are 02h, 21h, pid, 0, 0, 0, 0, 0, where pid is the PID code (see Table 7-7).
- Response is at ID 07EBh.
- Data bytes are number of bytes following with actual data, 61h, pid, up to 5 data bytes.
- Length is 8 data bytes.

TABLE 7-7 PID Support Messages

ID	Function	Len	Byte 3	Byte 4	Byte 5	Byte 6	Byte 7
F1h	State	6	State (1)	DOD (2)	Repor.SOC (3)	Capacity (17)	0
F2h	Pack data	7	Voltage (4)		Current (5)	CCL (6)	DCL (7)
F3h	Modules state	6	No. of mods (8)	No. of loads (9)	Bad mod # (10)	Bad mod cnt (11)	0
F4h	Mod. vtg range	7	Min vtg (12)	Min vtg no. (13)	Avg vtg (12)	Max vtg (12)	Max vtg no (13)
F5h	Mod. temp range	7	Min tmp (14)	Min tmp no. (13)	Avg tmp (14)	Max tmp (14)	Max tmp no (13)
F6h	Mod. number	3	Number (15)	0	0	0	0
F7h	Mod. details	5	Vtg (12)	Temper. (14)	Status (16)	0	0

Notes:

1. State of system:
   a. 0: Fault
   b. 3: Ready—charge sustain
   c. 4: Ready—charge deplete
   d. 9: Plugged—off
   e. 10: Plugged—charging
   f. 15: Ready and plugged in
2. Depth of discharge (%) (0 = full)
3. SOC reported to vehicle (%)
4. Total pack voltage, 2 bytes (V) (0100h = 256 V)
5. Pack current (A), signed (0 = 0 A; 10h = 16 A; F0 = –16A)
6. Charge current limit (A): maximum acceptable current
7. Discharge current limit (A): maximum current available
8. Number of modules reporting, 0 to 120, normally 60 (PHEV-15) or 120 (PHEV-30)
9. Number of modules whose balancing load is on (0 to 120)
10. Number of last module that failed reporting (1 to 120)
11. Count of how many times a module has failed reporting (up to 255)
12. Module voltage (10 mV + 2.5 V) (0 = 2.5 V; 1 = 2.51 V, etc., FFh = 5.05 V)
13. Number of the module with that reading (1 to 120)
14. Module temperature (°C) (0 = 0°C, 1 = 1°C, FFh = –1°C)
15. Number of the module whose data are being shown while scanning all the modules
16. Module status. Bit 0: 1 = load is on; bit 1: 1 = module is bad
17. Pack capacity [Ah]

Conclusion

In simple terms, computers think in binary logic: 1s and 0s, on and off, yes and no.

Rather than using complex feedback systems that home in on the results you want (engines), you can implement a simple logic approach that moves you there directly (controller).

Also, as a result of improving technology, you can save on components while not compromising on safety and dollars. Cell phone and camera makers, who have raised miniaturization to an art form, were ecstatic about the technology and have been improving on every new generation. What that means for PHEV converters is smaller and less expensive electronics. So, there is a great future for PHEVs and PHEV conversions today. Going forward with each new generation of technology improves our chances for electric drive.

CHAPTER 8

Batteries

More than 85 percent of an average car's gasoline energy is thrown away as heat.
—Dr. Paul MacCready, *Discover*, March 1992

Today's batteries, motors, and controllers are all superior to their counterparts of decades ago. Contrary to those who say that a different type of battery will be needed before EVs are suitable at all, today's conventional lead-acid batteries of the deep-discharge variety are perfectly adequate for your PHEV conversion.

Figure 8-1 shows you an array of cells in a Ford Escape PHEV. This picture shows the two layers of cells, separated, with the upper layer having been removed and turned upside down. Note that the electronics are in the middle of the layer. All indications are that these electronics provide insulation, and that therefore all the wires coming out of the cell pack (other than the high-voltage wires) are at low voltage.

In this chapter, you'll learn about how batteries work and the language used to discuss them. You'll also be introduced to the different battery types and their advantages and disadvantages. Then we'll look at the best type of battery for your PHEV conversion today, the lead-acid type used in Chapter 10's conversion, and also look at probable future battery developments.

Battery Overview

A battery is a chemical factory that transforms chemical energy into electric energy.

Your PHEV's chassis involved mechanical aspects, and its motors and controllers dealt with electrical ones. While there are all sorts of battery developments going on today, this chapter will give you a brief background on battery development,

Figure 8-1 Array of battery cells in a PHEV conversion. Source: Wikipedia.

and give information about PHEV conversions and the different types of batteries for you to choose among.

Active Materials

The active materials are defined as electrochemical couples. This means that one of the active materials, the positive pole or anode, is electron-deficient, and the other active material, the negative pole or cathode, is electron-rich. The active materials are usually solid (lead-acid) but can be liquid (sodium-sulfur) or gaseous (zinc-air, aluminum-air). Table 8-1 gives a snapshot comparison of a few of these elements.

Table 8-1 Electrolytes

Federal Urban Driving Schedule (FUDS) cycle	Cycle 1	Cycle 2	Cycle 3	Cycle 4	Cycle 5
km	12	12	12	12	12
L/100 km (mpg)	1.6 (148)	1.19 (200)	1.27 (187)	3.2 (74)	3.6 (66)
Wh/km	77	80	78	19	10
Engine times on	9	7	7	24	25
Highway cycle					
km	16.4	16.4	16.4	16.4	16.4
L/100 km (mpg)	2.12 (112)	1.95 (122)	2.22 (107)	3.3 (72)	3.8 (62)
Wh/km	62	61	57	14	0

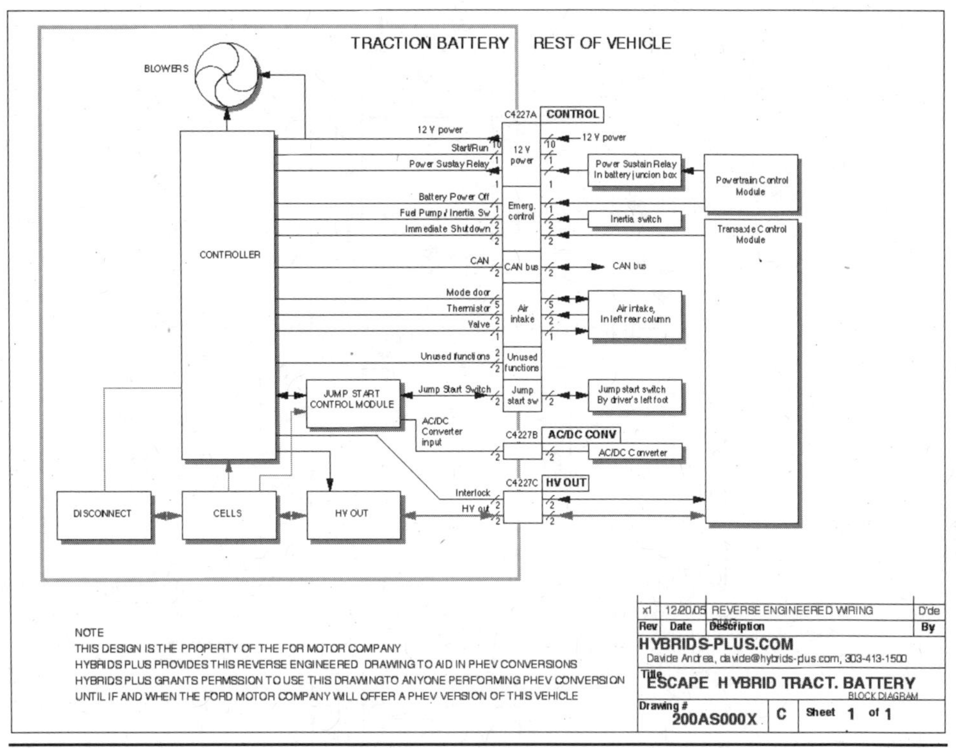

FIGURE 8-2 Ford Escape hybrid traction battery diagram. Source: EAA-PHEV.

The electrolyte provides a path for electron migration between electrodes, and also participates in the chemical reaction in some cells. The electrolyte is usually a liquid (an acid, salt, or alkali added to water), but it can be in jelly or paste form.[1]

Overall Chemical Reaction

Combining the active materials into compounds that further combine with the action of the electrolyte significantly alters their native properties. The true operation of any battery is best described by the chemical equation that defines its operation.

In a charged lead-acid battery, the positive anode plate is nearly all lead peroxide ($Pb0_2$), its negative cathode plate is nearly all sponge lead (Pb), and its electrolyte is mostly sulfuric acid (H_2SO_4). (See the top of Figure 8-3.) When the battery is in a discharged condition, both plates are mostly lead sulfate ($PbSO_4$), and the acid electrolyte solution used in forming the lead sulfate becomes mostly water (H_2O). (See the bottom of Figure 8-3.)

Discharging Chemical Reaction

The general equation gives a more accurate view when it is analyzed separately at each electrode. When discharging, the cathode acquires the sulfate (SO_4) radical

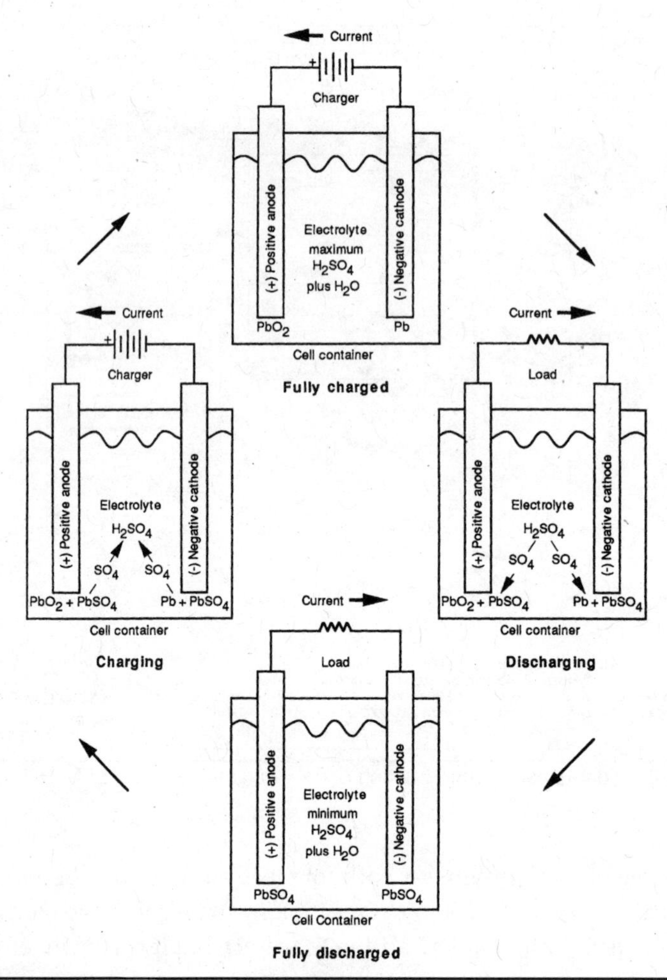

FIGURE 8-3 Chemistry of a simple lead-acid cell.

from the electrolyte solution and releases two electrons in the process. These electrons are acquired by the electron-deficient anode. While discharging (at the right of Figure 8-3), both electrodes become coated with lead sulfate ($PbSO_4$), and the sulfate (SO_4) radicals in the electrolyte are consumed. Before all the sulfate (SO_4) radicals in the electrolyte are consumed, the battery is fully discharged.

Charging Chemical Reaction

The charging process (at the left of Figure 8-3) reverses the electron flow through the battery and causes the chemical bond between the lead (Pb) and the sulfate ($S0_4$) radicals to be broken, releasing the sulfate radicals into the solution. When all the sulfate radicals are again in solution with the electrolyte, the battery is said to be fully charged.

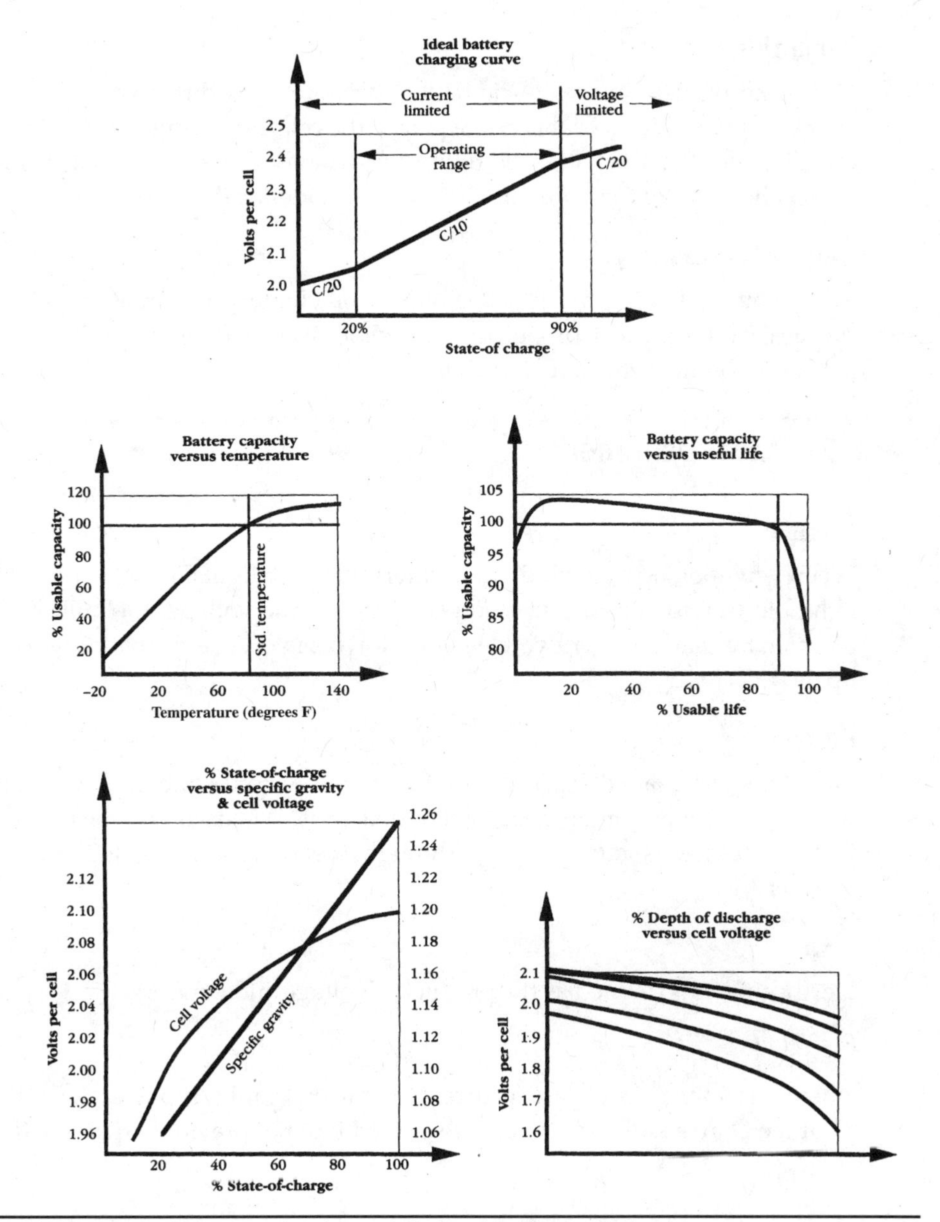

FIGURE 8-4 Lead-acid battery charging and discharging characteristics.

State of Charge

Voltage can be used to determine a battery's state of charge. Since a hydrometer, the device used to measure specific gravity, can introduce inaccuracy and contaminate a battery's cells, state of charge is determined electronically.

Equalizing

Over time, the cells in a lead-acid battery begin to show differences in their state of charge. The only cure for this is to equalize the cells by raising the charging voltage even higher after the battery is fully charged and maintaining it at this level for several hours until the different cells again test identical.

Outside Your Battery

When you hook up to the closed container of a battery, it exhibits certain external physical and electrical properties. You should be relatively familiar with these properties because they are useful to you.

Basic Electrical Definitions

Voltage

When you hook up a lightbulb to a battery, the bulb lights up. When you hook the lightbulb to two batteries in a series to double the voltage, the bulb lights even brighter. Similarly, battery voltage goes down as you use the battery—as you use up its capacity.

Current

The current (the rate of electron flow) corresponds to the rate of flow of the water coming out the bottom of a jug with a hole in it. When you doubled the voltage, you sent twice as much current through the wire, and the lightbulb became brighter.

Power

Electric power is defined as the product of voltage and current:

$$P = VI$$

where V is voltage in volts, I is current in amperes, and P is power in watts.

If the Ohm's Law equation is substituted into the previous equation, then

$$P = I^2R$$

This equation defines the power losses in the resistances in the circuit—either an external load or an internal battery.

Efficiency

Battery efficiency is

Efficiency = power out/power in

Battery Capacity and Rating

Capacity and rating are the two principal battery-specifying factors. Capacity is the measurement of how much energy the battery can contain, analogous to the amount of water in the jug. Capacity depends on many factors, the most important of which are

- Area or physical size of the plates in contact with the electrolyte
- Weight and amount of material in the plates
- Number of plates and type of separators between plates
- Quantity and specific gravity of the electrolyte
- Age of the battery
- Cell condition—sulfation, sediment in the bottom, and so on
- Temperature
- Low-voltage limit
- Discharge rate

Battery capacity is specified in ampere-hours. In practical terms, a battery with a capacity of 100 ampere-hours can deliver 1 ampere for 100 hours (known as a C/100 rate), but would not necessarily be able to deliver the much higher 100 amperes for 1 hour (known as a C/1 rate).

Power Density (Orgravimetric Power Density)

Also known as *speck power*, this is the amount of power that is available from a battery at any time (under optimal conditions), measured in watts per pound of battery weight. It translates directly into the acceleration and top speed performance your EV can get from its batteries.

Energy Density

This is the amount of power that is available from a battery for a certain length of time (under optimal conditions), measured in watt-hours per pound of battery weight. It translates directly into the range your PHEV can get from its batteries.

Volumetric Power Density

This is power density measured in watts per gallon or watts per cubic foot—volume rather than weight.

Battery Performance

Battery capacity is also highly dependent on the age of the battery. One observation is that a brand new EV battery pack will not give you as good a result as one that's been used for a while. Another observation is that once you begin to see battery performance go down significantly, it's time to think about buying another set of batteries.

The best way to determine the battery performance is with a digital voltmeter. Figure 8-5 shows a voltmeter and voltage recorder. With a voltmeter, you will be able to observe both voltage levels and current levels, and/or you can have the voltmeter readout drive the battery charger electronics directly. You can even monitor the voltages of individual cells if your battery type has external cell straps, all without the trouble of opening your batteries, dealing with sulfuric acid and hydrometers, and so on.

Other Factors Regarding the PHEV Battery

- *Battery construction.* From a manufacturing viewpoint, a lead-acid battery is one of the most efficient things going. More than 97 percent of all batteries are recycled, and 100 percent of every battery is recyclable.

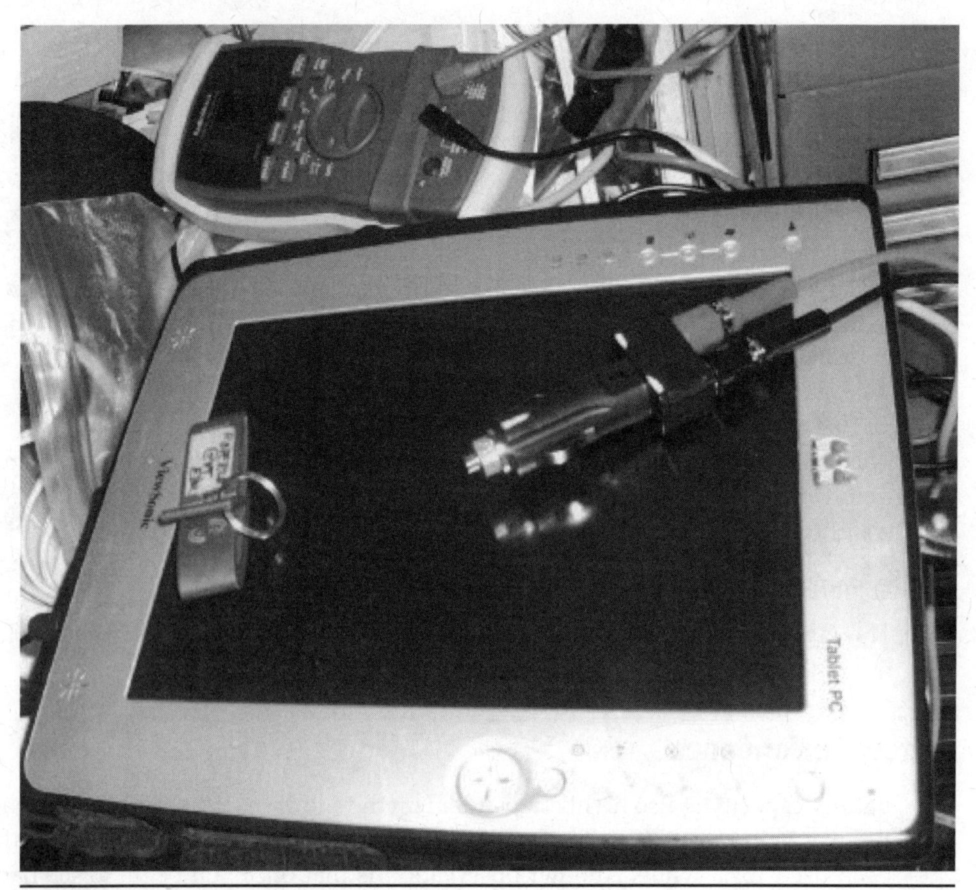

FIGURE 8-5 Voltmeter and voltage recorder.

- *Plates*. Battery plates are formed on a wirelike grid of lead alloy (antimony is sometimes used to stiffen the lead); a mudlike lead oxide, sulfuric acid, and water paste is applied to them and allowed to harden.
- *Case or container.* The case is a plastic or hard rubber one-piece rectangular container with three or six cells molded into it.
- *Cell connectors or links.* These connectors can be inside the battery (through or over the cell partitions) or outside the battery via the link connector (see Figure 8-6).
- *Terminal posts.* These might be taper top, side terminal, or L-type on starting batteries, but they are usually of the stud type on deep-discharge batteries (a heavy-duty post with a bolt and a washer).

Battery Installation and Maintenance Guidance

If you think of what you will be doing with your batteries, it will help you during the installation process. Tender loving care and maintenance of your batteries will reward you many times over. All it really takes is a plan, a schedule, and the discipline to do it. The plan starts with a notebook or logbook. The data you need are the voltage from a battery (or two or three—not the whole pack), the odometer reading, the date, and a comments section listing the amount of water you added and anything unusual you noticed. The schedule is weekly or biweekly. In general, here are your areas for concern and maintenance.

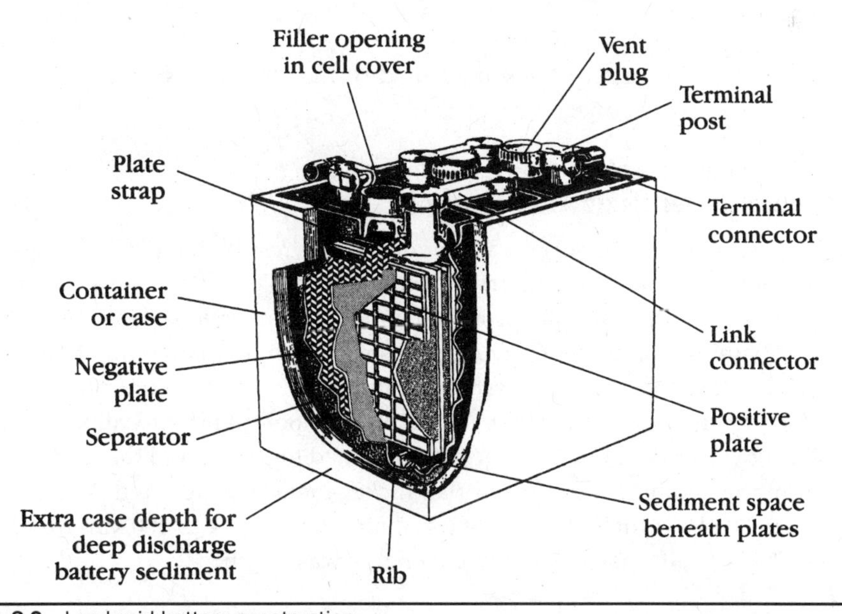

FIGURE 8-6 Lead-acid battery construction.

Safety

- Potentially lethal voltages are involved. It is important that the high-voltage wiring be done by an electrician or an engineer who is experienced in high-voltage safety.
  - Once the conversion is complete, all high voltages are inside screw- or bolt-secured areas, but these areas are exposed during parts of the conversion process, during battery replacement and other servicing, and possibly after a crash.
  - In keeping with hybrid automotive standards, high-voltage cabling is labeled with orange. (As no. 4 gauge and larger orange wire is not readily available, we specify the addition of orange shrink-wrap at each end.)
- This design should be adequate to contain the batteries under any normal driving conditions, but extreme or off-road maneuvers could damage the installation, potentially causing a hazard.
- More importantly, although we believe the parts are well secured, we are not automotive design engineers, the design has not been crash-tested, and its characteristics during and after a crash are unknown. In particular, it is uncertain whether the battery box would remain intact and in place during a rollover incident.
  - Lead-acid modules could tear out of their brackets and fly around the passenger compartment.
  - Although these AGM lead-acid modules are not flooded, they could leak acid if they were crushed.
  - Short circuits could arise, causing sparks and hot, molten metal, and possibly igniting a fire.
  - The battery pack may modify the characteristics of the vehicle's rear crush zone.[2]

Dynamometer Testing of PHEV Conversion Prius Vehicles Using A123 Systems

The most extensive set of dynamometer test results for PHEVs was done at the Argonne National Laboratory (see Table 8-2). The data in this paper that are of particular interest are those for the Hymotion conversion of the Toyota Prius using A123 lithium ion batteries. During this test, though, many limitations were placed on the Prius conversion. First, the modifications of the controller software were restricted and the power that came from the motor and the batteries was reduced. The power from the A123 battery was limited to 10 kW, and the total electric power from the combination of the Prius nickel–metal hydride and A123 batteries was about 20 kW. A fuel economy of less than 100 mpg (2.4 L/100 km) was experienced in Davis, California, even after the engine was warm.[3]

TABLE 8-2 ANL Dynamometer Test Results for the Hymotion PHEV Prius Using A123 Batteries

Type of hybrid driveline	System voltage V	Usable energy storage	Maximum pulse power at 90% efficiency kW	Cycle life (number of cycles)	Usable depth-of-discharge
Plug-in	300–400	6–12 kWh	35–70	2500–3500	Deep 60–80%
Charge sustaining	150–200	100–150 Wh	25–35	300K–500K	Shallow 5–10%
Micro-hybrid	45	30–50 Wh	5–10	300K–500K	Shallow 5–10%

Source: Andrew Burke and Eric Van Gelder, *Plug-in Hybrid-Electric Vehicle Powertrain Design and Control Strategy Options and Simulation Results with Lithium-Ion Batteries*, Institute of Transportation Studies, University of California–Davis, One Shields Ave., EET-2008, European Ele-Drive Conference, International Advanced Mobility Forum, Geneva, Switzerland, March 11–13, 2008.

Today's Best Battery Solution

You already know that this book recommends lithium ion batteries as the best solution for today's PHEV converters. You also know what type of lead-acid battery to buy and a lot about its characteristics. Your choice is made even easier because there are only a certain number of battery vendors in your immediate geographic area to choose among. Unlike in buying motors, controllers, and other parts, you're not likely to be ordering your batteries by mail. Your choice basically comes down to who offers the best price on the batteries you want, and what capacity, rating, voltage, size, and weight you need. In a slight departure from the previous chapters, we're going to recommend one manufacturer, then look at several alternative offerings from its line to give you the flavor of the real choices you will encounter.

A123 has developed a high-energy cell based on its mass-produced Nanophosphate chemistry that provides higher energy density than the M1 power tool cells, a trait that is critical for energy-dependent PHEV applications. While more energy dense, the M1HD products still retain high power capability, which is critical for charge-sustaining operation in PHEV applications (see Figure 8-7).

In the study results given in Table 8-3, all the vehicles utilized lithium ion batteries, with characteristics typical of the A123 cells: energy density: 88 Wh/kg; resistance: 0.025 ohm-Ah; pulse power density at 95 percent efficiency: 730 W/kg.

Table 8-3 Energy Storage Requirements for Various Types of Hybrid-Electric Vehicles

Chemistry	Best or Worst	Usable Wh/kg	Cycle life	Yr daily driving	$/usable kWh	$/kWh thruput	Cents/ EV-mi	kWh	$	EV mi	Wt, lb
PbA (current)		16	400	1.1	$380	$0.95	20.0	2.1	$798	10	289
NiMH	Worst	36	2000	5.5	$1,200	$0.60	12.6	4.2	$5,040	20	257
NiMH	Best	36	4000	11.0	$800	$0.20	4.2	4.2	$3,360	20	257
Li-ion	Worst	56	1000	2.7	$1,200	$1.20	25.2	4.2	$5,040	20	165
Li-ion	Best	100	4000	11.0	$800	$0.20	4.2	6.3	$5,040	30	139
NiZn	Worst	36	500	1.4	$500	$1.00	21.0	4.2	$2,100	20	257
NiZn	Best	36	2000	5.5	$350	$0.18	3.7	4.2	$1,470	20	257
Firefly PbA	Worst	36	1000	2.7	$350	$0.35	7.4	4.2	$1,470	20	257
Firefly PbA	Best	45	4000	11.0	$250	$0.06	1.3	5.25	$1,313	25	257

Andrew Burke and Eric Van Gelder, *Plug-in Hybrid-Electric Vehicle Powertrain Design and Control Strategy Options and Simulation Results with Lithium-ion Batteries*, Institute of Transportation Studies, University of California-Davis, One Shields Ave., EET-2008, European Ele-Drive Conference, International Advanced Mobility Forum, Geneva, Switzerland, March 11–13, 2008.

Figure 8-7 A123 Systems batteries for PHEV systems.

Tomorrow's Best Battery Solution—Today

USABC Continues to Come to the Rescue

Let's look at future battery trends, starting with the consortium that's pushing the outside of the battery envelope—the USABC.

In 1991, Ford, General Motors, and Chrysler (joined by the Department of Energy and the Electric Power Research Institute) had a better idea—the United States Advanced Battery Consortium (USABC). The Electric and Hybrid Vehicle Research, Development and Demonstration Act of 1976 recognized the need for battery development, the Department of Energy defined and funded it, and the USABC focused the efforts.

Also, on September 25, 2008, the USDOE announced that it is providing $17.2 million to further the development of advanced batteries, and another $2 million for the study of future plug-ins. Cost sharing with the USABC will allow up to $38 million in battery research and development.[4]

The near-term result was that a plethora of projects was honed down to just three high-energy battery research areas that could deliver significant vehicle range and power advantages: lithium polymer, lithium–metal sulfide, and nickel–metal hydride. Some of these batteries are considered the best of the best battery technologies on the market.

The Big Picture on Batteries

Table 8-4, adapted from an SAE paper, shows the entire story at a glance. Notice that there are five different battery technologies on the list, and they are not equal. In very general terms, higher energy density and power density are desirable and look easy to do—on paper. Getting both at the same time, along with high cycle life and low cost in a battery that operates efficiently over a range of temperatures, can be manufactured and supported by infrastructure, and causes no harm to people or the environment, has proven to be a bit more elusive. Notice that none of the batteries developed thus far—even the tried and proved lead-acid battery—approaches its theoretical specific energy value. We still have a long way to go.

Battery Types[5]

Among rechargeables, there are lead-acid batteries and then there are all the rest. Let's look at all of them. Table 8-4 shows you most of the PHEV battery types and the respective trade-offs of each.

Table 8-4 Comparison of PHEV Battery Trade-Offs

Trojan Battery Model	Total kwh/lbs	Energy Decision Criteria	Total $/lbs	Cost Decision Criteria
T-105	21.34		1.237	
T-125	21.36	Highest 6-volt energy/lb	1.227	Lowest 6-volt cost/lb
T-145	20.62		1.669	
27TMH	23.40	Highest 12-volt energy/lb	1.247	Lowest 12-volt cost/lb
5SHP	23.02		1.795	

Lead-Acid Batteries

Don't expect this type of battery to go away soon, if at all. The big money involved in the lead-acid battery business will make the lead-acid batteries of the early 2000s as superior to today's as today's are to their 1980s counterparts. Along the way to higher specific energy and specific power, lead-acid batteries will also evolve variants like sealed batteries (higher cost and lower efficiency, but with the convenience of not having to add water), flow-through batteries (the conventional type you're accustomed to, but improved by greater plate thickness, improved separators, and higher specific gravity electrolyte solution), tubular batteries (electrode improvement), and gelled batteries (electrolyte improvement).

As mentioned on the EAA-PHEV wiki site, lead-acid batteries add 300 or more pounds to the conversion weight of the vehicle and provide 10 miles of electric range per charge (16.7 usable Wh/kg).

- Operating costs are high because these batteries have an expected cycle life of only 300 to 400 deep cycles, thus providing only one to two years of daily driving. At 400 cycles, 10 electric miles per 2.1-kWh cycle, and $800/pack, battery cost is $0.95/kWh throughput or $0.20/electric mile (in addition to the cost of electricity, which is usually 2 to 4 cents/mile depending on utility rates).
- For decent battery life, the battery must always be charged within a day of discharge, making charging a required rather than an optional operation (if you are planning to drive somewhere without access to electricity, temporarily turn off PHEV operation).
- Lead-acid batteries perform very poorly in cold weather. The design includes a thermally insulated battery pack that is heated during charging, this feature has been insufficiently tested as a result of the moderate California temperatures during development.[6]

Nickel-Cadmium Batteries

Nickel-cadmium (NiCad) batteries are the type you use in your portable computer, shaver, or appliance. They are unquestionably better than lead-acid batteries in terms of their ability to deliver twice as much energy pound for pound; they also have about 50 percent longer cycles. But the nickel-cadmium electrochemical couple delivers a far lower voltage per cell (1.25 volts), meaning that you need more cells to get the same voltage. They are also far more expensive (four times as much and up). There are fewer sources for the heavy-duty EV-application batteries, and cadmium itself is harder to obtain and has generated environmental concerns. Finally, most of the nickel-cadmium technology development is taking place overseas (in England, France, Germany, and Japan).

Close on the heels of lead-acid today, this battery type promises to be even better in the future. Its advantages over lead-acid today (a less pronounced decrease in capacity under high-discharge currents, higher cycle life, slower self-discharge rate, better long-term storability, and improved low-temperature performance) will continue to motivate markets to fund development aimed at overcoming its disadvantages in relation to lead-acid: higher cost and environment-related cadmium issues.

Nickel-Iron Batteries

The "Edison battery" used in the early 1900s EVs is a poorer choice. These batteries offer a higher cycle life (about twice as many), deliver slightly more energy pound for pound (about a third more), and are very rugged mechanically. However, the nickel-iron electrochemical couple delivers only slightly more voltage per cell than a NiCad (about 1.3 volts) and has a high internal resistance and self-discharge rate (10 percent per week). Its performance degrades significantly with temperature (both above and below 78°F). These batteries are far more expensive (four times as much and up), there are few sources for them (they're made only in Europe and Japan), and there is little technology development taking place.

All the battery development going on in the labs (which we'll look at briefly later in the chapter) is great, but you can't buy these batteries now. Your choice boils down to the good old lead-acid battery. But lead-acid batteries are not all created equal. Confining our discussion to the PHEV applications of actual hybrid electric vehicles, you have three types to choose from and they are: nickel-metal hydride, lead-acid, and lithium based technologies. While seven battery technologies are listed, the most popular are the three mentioned before.

Nickel-Metal Hydride

This environmentally benign alter ego to the nickel-cadmium battery is also flat-out superior to it in specific energy and specific power comparisons, and it should become the preferred alkaline battery of the future. The USABC certainly thinks so and is investing its research and development dollars in it.

Sodium Sulfur

Yes, sodium is combustible in air. However, this continues to be one of the most promising advanced battery systems for PHEV propulsion. While many problems remain to be solved, with low-cost production being perhaps the largest among them, sodium sulfur is a hot technology in more ways than one.

Sodium Metal Chloride

Sodium metal chloride batteries have higher open-circuit voltages and better freeze/thaw and failure modes than sodium sulfur, so you have a real winner.

Lithium Polymer

This battery's specific energy and specific power numbers evoke nothing but envy from its competitors and mouth-watering anticipation from its advocates. The whole lithium group has shown great promise in the labs, and smaller lithium polymer batteries have greatly impressed users in the computer industry. But lithium still has to deliver on its promise when packaged in the giant, economy, suitable-for-powering-PHEVs size.

Lithium-Nanophosphate

Lithium iron phosphate (LFP) is a cathode material for lithium iron phosphate batteries that is getting attention from the industry. Valence Technologies sells the only large-format lithium iron phosphate battery currently available. The most important merits of this battery type are safety and high power. $LiFePO_4$ is one of three major compounds and technology in the LFP family. The other two are Nanophosphate and NanoCocrystallineOlivine.

Lithium Iron Disulfide

The promise of lithium-iron disulfide batteries on the high-temperature side is equally mouth-watering. This battery's specific energy numbers are the best of all, and its specific power numbers simply leave all others in the dust.

Lithium Ion

This has to be the most popular long-range battery on the market today. The advances in this technology make it the next best solution for the present and the future. The lithium ion moves from the anode to the cathode during discharge and from the cathode to the anode during recharging. It is extremely popular in consumer electronics and now power tools. It is light, has a slow energy degradation, and has no memory issues.

A123 Systems. A123 Systems of Watertown, Massachusetts, has developed an amazing lithium ion battery using a Nanophosphate chemistry that was coordinated with the U.S. Department of Energy that is being used in plug-in hybrid technologies from Hymotion (see Figures 8-8 and 8-9). It is also being used in hybrid electric transit buses (Orion buses) and in airplanes when the airplane is docked at an airport for power, and it will soon be in hybrid electric cars so that they can be even more efficient than they are today. An electric motorcycle called the KillaCycle that uses the A123 batteries was recently developed. Itn a race that was shown on the cable show *Planet Green*, it went 168 mph in 7.824 seconds. Figure 8-10 shows you

Plug-in+ PHEV Safety

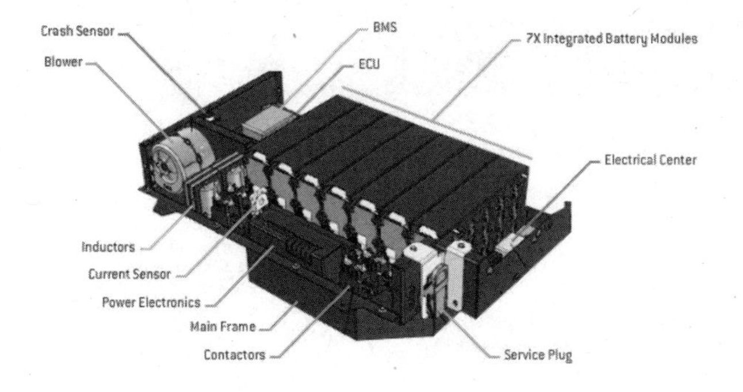

Nanophosphate™ Chemistry: the Foundation for Safe Systems

A123's Nanophosphate chemistry provides the foundation for a more higly abuse tolerant lithium ion solution in PHEV application. In addition, cell and system design play crucial roles in overall abuse tolerance.

High power systems require advanced design to manage the safety of the system over the life of the battery. Designing for safety is based upon intimate knowledge of the various aspects of electrode, cell and system design.

FIGURE 8-8 A123 Systems battery pack. Source: A123 Systems website.

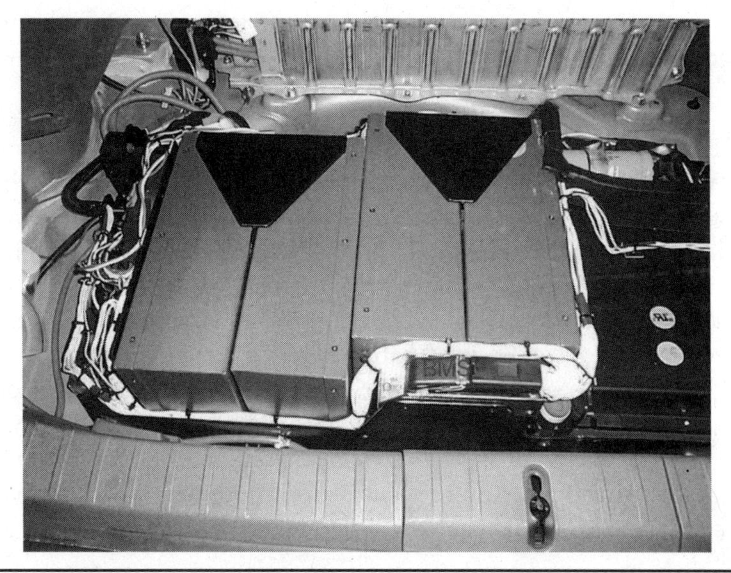

FIGURE 8-9 A123 Systems battery packs in Hymotion conversions.

Plug-in+ Hymotion L5 Plug-in Conversion Modules PHEV Solution

Commuting Distance	Measured City MPG	Measured Highway MPG
20 miles	174	117
40 miles	153	103
60 miles	124	90

Blended Mode PHEV(5kWh)
Li-Ion battery pack added to stock NiMH pack
Installed in Argonne's highly instrumented Prius

Argonne National Labs Validates Petroleum Savings with L5 PCMs

The technology of A123's Hymotion L5 Plug-in Conversion Modules have been validated at Argonne National Labs, and have shown greater than 100 miles per gallon in city and highway use. This key validation shows the type of impact that after-market Plug-in Conversion Modules can have on petroleum consumption. A123's foundational Nanophosphate™ technology is leveraged by Hymotion products to produce new PHEV solutions.

Figure 8-10 A123 Systems testing with the U.S. Department of Energy. Source: A123 Systems; http://www.a123systems.com.

the certified Argonne National Laboratory range that the A123 Systems battery pack provided to a Toyota Prius.

Conclusion: The Future Is Amazing!

PHEVs using lead-acid batteries are now available for $6,000 to $10,000, those using nickel-metal hydride for $8,000 and up, and those using lithium chemistries for $10,000 and up for pure electric vehicles. Whereas, for PHEVs the cost is about 50–60% of that. Conversions are mostly for the Prius, but there are a few for the Ford Escape/Mercury Mariner hybrid SUVs. At these prices, people are buying the "environmental feature"—they want to be *among the first owners of the world's cleanest extended-range vehicles*. They are early adopters, buying "Version 1.0" PHEVs with "Good Enough to Get Started" batteries.[7]

It may be possible to retrofit more advanced batteries to these conversions. This will probably require upgrading to CalCars' not-yet-designed next version of the logic board, and will also probably require additional battery management electronics. Any new battery's enclosure, mounting, and thermal management system will no doubt also be very different.[8]

Lithium Ion and Nickel-Metal Hydride Are the Current Battery Technology Success Stories, but Lead Is the Least Expensive to Get the Job Done Today

The cathodes of some early 2007 lithium ion batteries were made from lithium-cobalt metal oxide. This material is expensive, and cells made with it can release oxygen if they are overcharged. Interestingly enough, if you replace the cobalt with a lithium ion phosphate battery pack, the cells will no longer have the risk of burning or releasing oxygen under any charge.

Battery Charge for PHEVs

PHEVs usually require deeper charging and discharging cycles for the batteries. Conventional hybrids, since they do not have a pure electric vehicle component, do not need to cycle the batteries. Because the number of full cycles influences battery life, this may be shorter in PHEVs than in traditional HEVs, which do not deplete their batteries as fully.

However, Figure 8-11 shows how great the battery pack voltage vs. current really is. Notice that the current is generally within a specific range and is relatively consistent throughout. That is pretty much what you can expect from your PHEV!

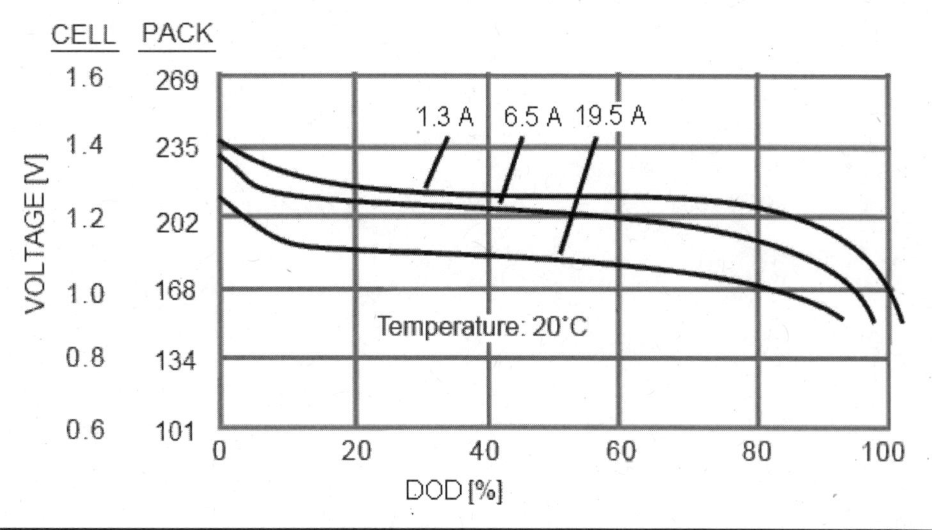

FIGURE 8-11 Prius battery pack voltage vs. current.

CHAPTER 9

The Charger and Electrical System

An efficient charger is an indispensable part of any plug-in hybrid electric vehicle.

The charger is an *attached* and *inseparable* part of every PHEV battery system. The way your batteries are discharged and recharged determines both their immediate efficiency and their ultimate longevity. As with motors, controllers, and batteries, technology has made today's chargers superior to their counterparts of a decade ago.

In this chapter, you'll learn about how chargers work and the different types, meet the best type of charger to choose for your PHEV conversion today (the conversion in Chapter 10), and look at likely future developments in chargers.

You'll also look at your PHEV's electrical system in detail and learn about its components, so that when you meet them again during the conversion process in Chapter 10, they will be familiar to you.

Charger Overview

Chapter 8 dealt with the discharging and charging of batteries. Now we'll look at charging the car. It's a wise business decision to invest a few hundred dollars in a battery charger that gets the most out of a battery pack that can cost a thousand dollars or more. A great, efficient charger is critical to any PHEV.

It's what makes it the electric vehicle it is.

The objective here is to give you a brief background and get you into some good chargers for your PHEV conversion so that there is less hassle. As I mentioned in *Build Your Own Electric Vehicle*, you have two battery charger choices today: off-board or on-board.

We'll look at both and give our recommendations. Let's start with a look at what goes on during the discharging and charging cycle for the lead-acid battery to understand what the battery charger has to do.

Battery Discharging and Charging Cycle

As you already know from Chapter 8, batteries behave differently during discharging and during charging—two entirely different chemical processes are taking place. Batteries also behave differently at different stages of the charging cycle. Let's start with a look at an actual battery, then look at the specifics of the discharging and charging cycle.

Capacity, cell voltage, and specific gravity all diminish over time as you use your battery packs. However, the packs diminish less over time if you use lithium ion or nickel-based technology.

The Ideal Battery Charger

The basic rule is: charge the battery as soon as it's empty, and fill it all the way up. The charging rate rule is: charge the battery more slowly at the beginning and end of the charging cycle (when it's below 20 percent and above 90 percent).

Manzanita Micro Moving from Electric to PHEV Too!

Ryan and Rich Rudman are developing conversion kits based on an enhanced PFC-30 from manzanitamicro.com.

The Manzanita Micro Is a Great Charger for PHEVs

The PFC-30 will permit flexibility in battery pack sizes from about 10 to 30 batteries, among other battery management and charge-rate enhancements. The first battery rack will be configured for 15 Hawker EP 26 or Hawker Areo AB26 SLA batteries.

Since the PFC charger is used as a DC-to-DC converter, it is possible to use a wide variety of battery pack configurations. Rich has also used battery packs of 20 and 24 smaller 20-ampere-hour (Ah) batteries similar to the EVP20-12B used by CalCars, and more recently Rich has also implemented a battery pack made up of 40-Ah Thunder Sky lithium ion modules. It is best if the add-on battery pack is of a higher voltage than the OEM battery to reduce switching losses and increase the efficiency and power out of the PFC charger.[1]

As shown in Figure 9-1, Manzanita Micro has developed a PHEV system for the Toyota Prius. Its charger is the green box on the left, where the high-voltage wiring is installed.[2]

Figure 9-2 is from Elithion and shows that the company has developed a low-cost, CAN bus, PFC (power-factor-corrected) charger for large battery packs that is simple to install and operate.[3]

Figure 9-1 Manzanita Micro revamped PHEV drive system—the charger is the box on the left.

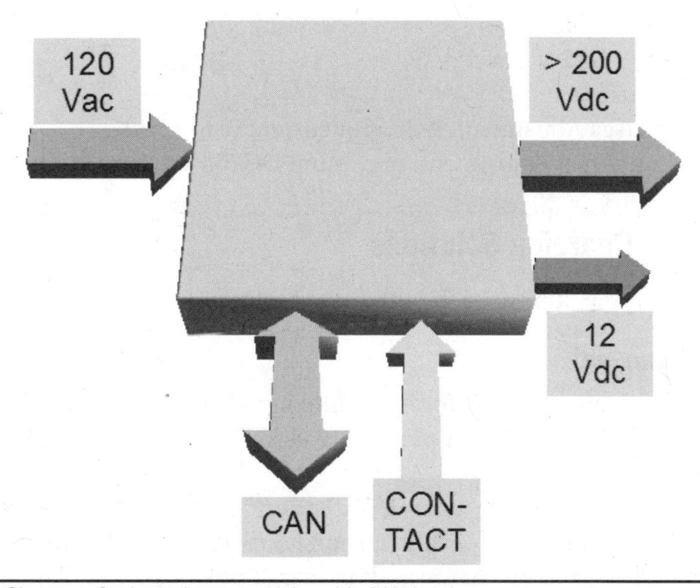

Figure 9-2 Chargers for large battery packs with CAN bus.

- Electrical:
  - Minimum battery voltage (at discharged battery): 190 V DC
  - Maximum battery voltage (at fully charged battery): adjustable from 200 to 450 V DC
  - Input current: 15 A AC, user adjustable down to 1 A AC
  - Maximum power (at 120 V AC): 1,800 W
  - Efficiency: > 95%
  - Power factor: > 0.99
  - Charge current shape: fully rectified sine wave
  - Optional 13.5-V, 1-A output to float charge 12-V lead-acid auxiliary battery
- Mechanical and environmental:
  - Temperature range: –20 to +80°C
  - Dimensions: 3" × 5" × 7"
- Connectors:
  - AC input connector: IEC 320-C14
  - DC output connector: Anderson 15-A powerpole
  - Control connector: six-pin Molex Mini-fit JR connector: Common, 12 V, control
    - 1: GND
    - 2: CAN-lo
    - 3: CAN-hi
    - 4: Control in [closed contact to GND = on, open contact (1 kohm pull-up to +12 V) = off]
    - 5: N.c.
    - 6: 12 V out, 1 A
- Adjustments:
  - User adjustment: AC input current (0 to 15 A)
  - Internal adjustment: maximum DC out voltage (200 to 450 V)

Other Battery-Charging Solutions

Let's take a brief look at each of the other solutions.

Rapid Charging[4]

The Japanese are the most forward-thinking in this area. However, Shai Agassi in Project Better Place, with his plan to develop EV infrastructure around the globe, is starting to get it, too.

In short, if charging your EV's battery pack in eight hours is good, then accomplishing the same result in four hours is better. Fortunately, you can use pulsed DC, alternating charge and discharge pulses, or just plain high-level DC to accomplish these results.

Unfortunately, you need to start with at least 240-volt AC (240-V three-phase is better), and you will probably overheat and shorten the life of any of today's lead-acid batteries in the process. However, if your lead-acid batteries are designed to accommodate this feature (a larger number of thinner lead electrodes, special separators, and electrolyte), it becomes easy. Even nickel-cadmium batteries can be adapted to the process if your wallet is bigger. You'll hear more about this idea as time goes by. (Three-phase 240-V will supply 50 percent more power than 480-V single-phase. Because adding one wire gives you three times the power, the world's power grid is three-phase.)

In a residential area, blocks are rotated between phases A, B, and C. Full-wave rectification of three-phase gives you DC with very little ripple. (The second, third, and fourth harmonics cancel.) In electric vehicles, three-phase motors can be reversed by swapping any two phases.

Replacement Battery Packs

This approach, with neighborhood "energy stations" replacing gasoline stations, will also have a role in the future. Future EV designs could be standardized, with underbody pallet-mounted battery packs. You would wheel into the energy station, drop the old battery pack, raise the new one into place and latch it down, pay by credit card (probably a deposit for the pack plus the energy cost for the charge), and be on your way within a few minutes.

Beyond Tomorrow—V2G

Serious infrastructure development is necessary for the real prize: roadway-powered PHEVs. Numerous papers and articles have been written about this idea, all because of its mouth-watering appeal. A simple battery pack powered by a lead-acid battery can give a regular gas car or a hybrid more than enough range to carry you to the nearest interstate highway. Once you're there, you punch a button on the dashboard. An inductive pickup on your EV draws energy from the roadway through a "metering box" in your vehicle (so that the utility company knows who and how much to charge), and you get recharged on your way to your destination without any pollution, noise, soot, or odors.

If the roadway and your EV are both of the "smart" design, you can even get travel information (weather, directions, and so on) while the roadway guides your vehicle in a hands-off mode and you read the morning newspaper or scan the evening TV news. All it takes is infrastructure development—read that as M-O-N-E-Y.

One way to do this is to allow charge control, smart charging, and Vehicle to Grid (V2G).[5]

V2G involves letting the power company decide when you should charge (or discharge to the grid). Smart Charge and V2G are charge control technologies that let the power company control a plug-in vehicle's charging or discharging in real time to prevent a plug-in vehicle from charging during a time of peak demand, or to buy back the energy from a plug-in vehicle's battery. It also enables both the owner and the power company to track the plug-in vehicle's usage and performance, both while on the road and while charging.

What's the Point of Charge Control?

During times of peak demand, the power company prefers to reduce the demand rather than turning on additional (dirtier and more expensive) power plants. In places that subscribe to such services, the power company shuts off appliances such as air conditioners until the peak demand period is over. Similarly, Smart Charge allows the power company to postpone charging until the peak demand period is over. Smart Charge is not as far-reaching as V2G (a technology that allows the power company to buy energy back from a plug-in car), but it is very effective because it is less expensive than V2G.

The hardware used in charge control (both V2G and Smart Charge) also monitors the vehicle (both when you are driving and when it is plugged in). As a side benefit, that information is available, through the Web, to both the power company and the owner, for the purpose of tracking and optimizing the performance of the vehicle.

At this point, there is no financial benefit from using charge control. However, one day the power utility may sell electricity to charge Smart Charge vehicles at a lower price, and may buy that electricity back at a premium.

What's the difference between Smart Charge and V2G?

In both V2G and Smart Charge, the power utility can control the power flow between a plug-in car and the power grid. However, in Smart Charge, power flows only from the grid to the car, while in V2G, power can flow in both directions.

Figure 9-3 shows how Smart Charge works, allowing your PHEV to be plugged into the grid and allowing the utility to manage when your vehicle can be charged.

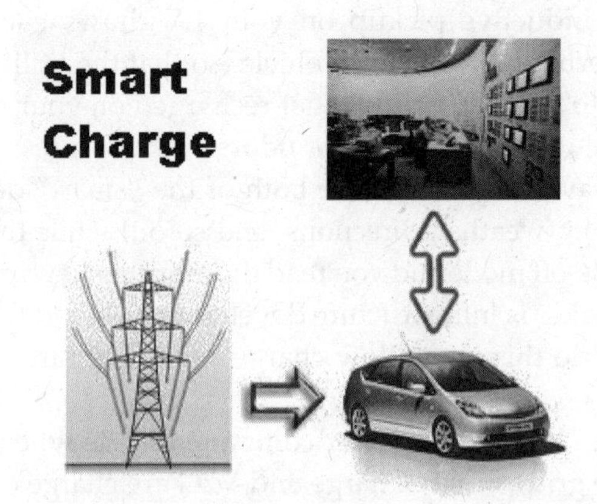

Figure 9-3 Smart Charge charge control system diagram. Source: Elithion.

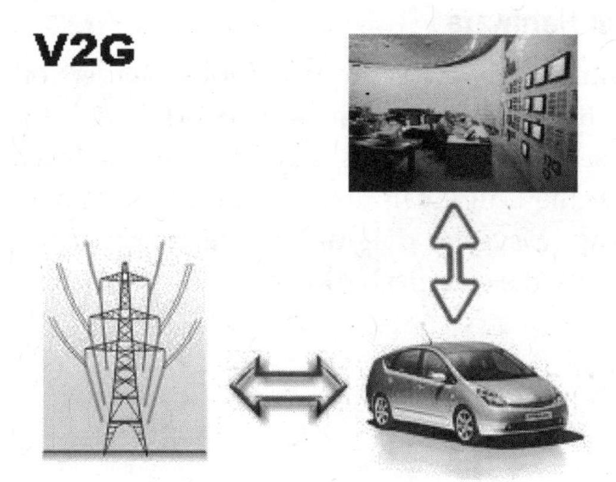

Figure 9-4 V2G charge control system diagram. Source: Elithion.

Thus, if the utility needs power and your PHEV has enough of a charge, it can decide not to charge your vehicle. Figure 9-4 shows that with V2G, the utility could take the power from your battery pack, rather like what happens when solar panels store energy and the utility can take your electric meter and reverse it to decrease your energy costs.

Charge control (either Smart Charge or V2G) could be used on a fleet of plug-in vehicles fitted with charge control hardware. Figure 9-5 gives a diagram of how the grid would communicate with the PHEV system inside the car.

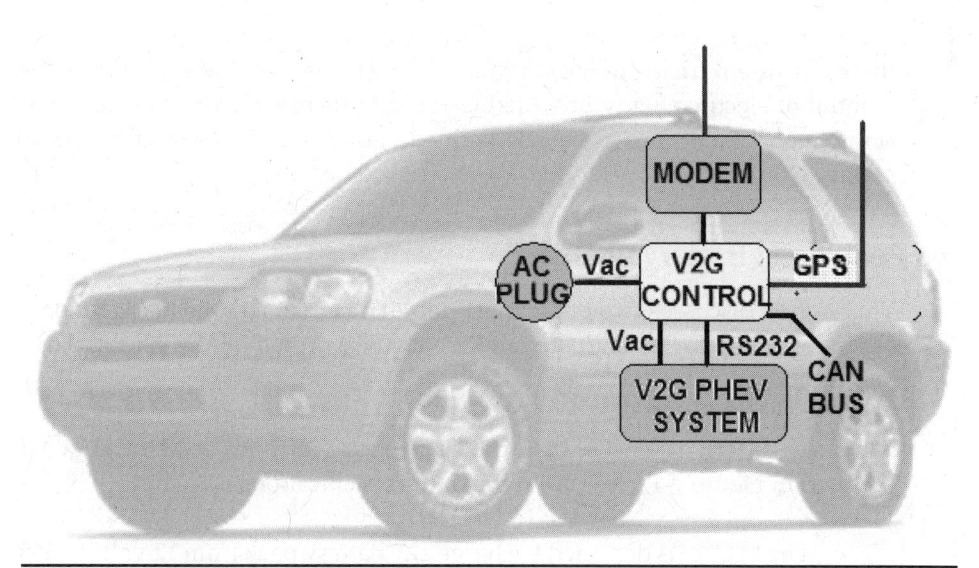

Figure 9-5 Vehicle diagram. Source: Elithion.

Charge Control Hardware

The charge control hardware is connected to the servers of a company that manages charge control through the cell phone network and the Internet. The owners, the utilities, and research organizations have access to the data from each plug-in vehicle.

In addition, the utility can disable charging or (in V2G) request discharging. Figure 9-6 shows a system diagram of the charge control hardware and how it interconnects with the electrical grid.

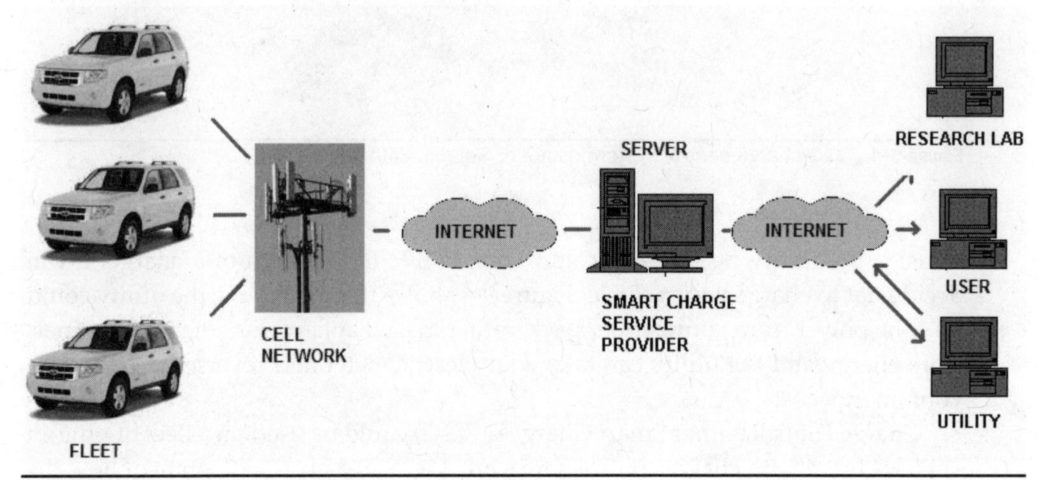

FIGURE 9-6 System diagram of charge control hardware.

V2G Reports That Can Be Generated for Consumers

Figure 9-7 could be the future of vehicle/electrical grid interconnections. Look at the drop in electricity usage on Wednesday in the sample report and how the amount of electric energy provided to the grid increased. This is an example of a status report for a plug-in vehicle with charge control that would be available in real time through the Web.

The Real-World Battery Charger

This section discusses two of the most popular chargers on the market today. They are the standard for the industry and are really accepted by the marketplace.

The Manzanita Micro PFC-20

Some say, "More expensive than the Zivan, but worth it." A Manzanita Micro is pictured in Figure 9-1. Here are some of the specifications:

- The PFC-20 is designed to charge any battery pack from 12 volts to 360 volts nominal (14.4 to 450 V peak). It is power factor–corrected and designed to

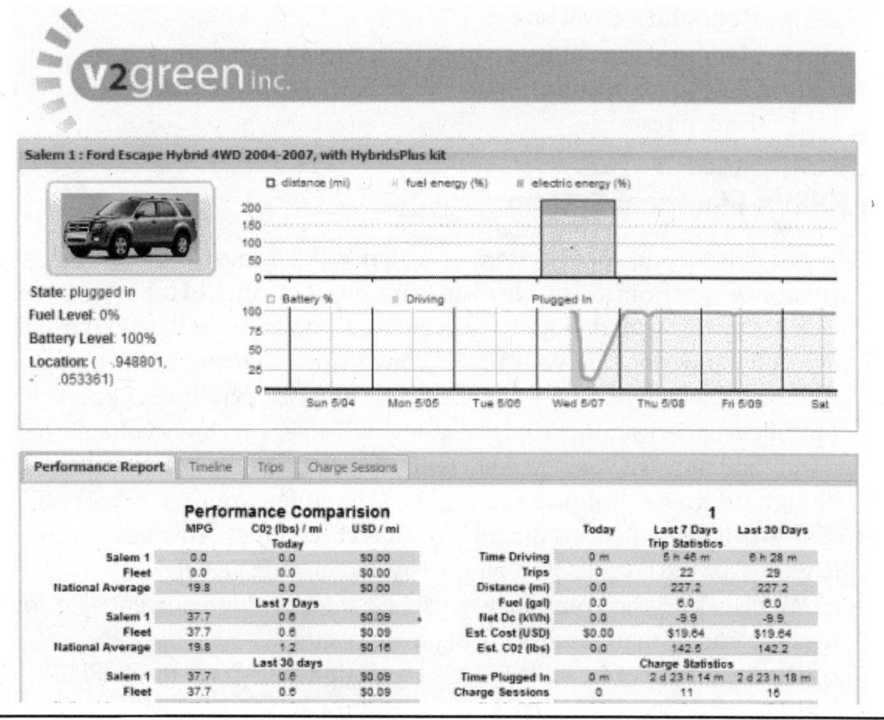

FIGURE 9-7 Example of a status report for a plug-in vehicle with charge control, available in real time through the Web. (Source: http://liionbms.com/php/about_charge_control.php)

either put out 20 amperes (if the battery voltage is lower than the input voltage) or draw 20 amperes from the line (if the line voltage is lower than the battery voltage). The buck enhancement option on the PFC-20 will enhance the output to 30 amperes. There is a programmable timer that will shut off the charger after a period of time set by the user.

- For installation instructions, go to www.manzanitamicro.com/installpfc20revCnophotos.doc.

Elithion Charger

The Elithion[6] charger is significantly less expensive than other traction battery chargers on the market. Here are some reasons why it is a high-quality charger:

- Power-factor-corrected (PFC) input
- Nonisolated
- Fan cooled
- Regulation: input AC current or maximum output DC voltage, whichever is limiting
- Fuse protected: reverse battery polarity, output short, short to chassis
- Control: closed contact or CAN bus

- Monitoring: CAN bus
- Mandated regulatory testing: FCC Part 15 tested
- Optional regulatory testing (UL, CSA, CE, and so on): may be performed by the end user

Your PHEV's Electrical System

To say that the electrical system of a PHEV is its most important part is not an oxymoron—far from it. The idea is to leave intact as much of the internal combustion vehicle's instrumentation wiring as you need, and to carefully add the high-voltage, high-current wiring that your PHEV conversion requires.

This section will cover the electrical system that interconnects the motor, controller, batteries, and charger along with its key high-voltage, high-current power and low-voltage, low-current instrumentation components. Tables 9-1, through 9-4 give simplified and detailed drawings from the EAA-PHEV website. Table 9-1 shows a detailed diagram from HybridsPlus on the new electrical system diagrams for the Ford Escape plug-in hybrid electric vehicle.

We'll look at the components that go into the high-current and low-current sides separately, then discuss wiring it all together.

The heavier lines in Table 9-2 are Hymotion's diagram for its PHEV. Tables 9-3 and 9-4 show the EAA-PHEV Prius control board schematic and high-power schematic. Note the high-current connections for the PHEV. When you put a new motor, controller, battery, and charger together in your vehicle, you need contactor(s), circuit breakers, and fuses to switch the heavy currents involved. Let's take a closer look at these high-current components.

Main Contactor

A contactor works just like a relay. Its heavy-duty contacts (typically rated at 150 to 250 amperes continuous) allow you to control heavy currents with a low-level voltage. A single-pole, open main contactor is placed in the high-current circuit between the battery and the controller and motor, as shown in Figure 9-8. When you juice it up—typically by turning the ignition key switch on—high-current power is made available to the controller and the motor.

Figure 9-9 shows the universal case for the Manzanita Micro battery system. The contactor box is on the left side of the battery box and is seen in the figure. The Hawker box uses a separate contactor box that is mounted to the right of the Hawker box in the right rear fender.[7]

Main Circuit Breaker

A circuit breaker is like a switch and a resetting fuse. The switch plate and mounting hardware are useful—the big letters immediately inform casual users of your PHEV of the function of the circuit breaker.

TABLE 9-1 Detailed Diagrams of the New Electrical System for the Ford Escape Plug-In Hybrid Electric Vehicle

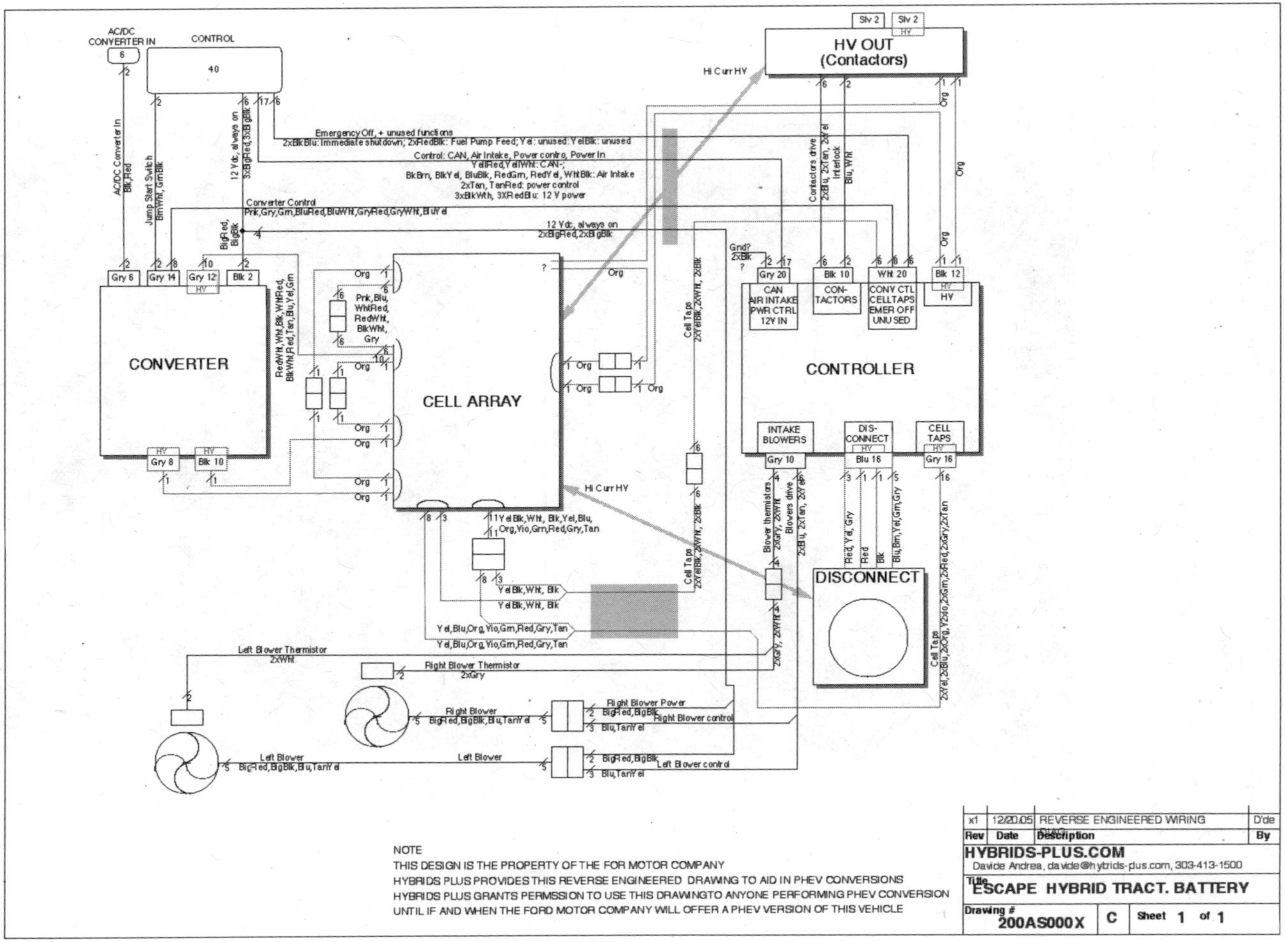

HV OUT (Contactors)

CONVERTER

CELL ARRAY

CONTROLLER

DISCONNECT

NOTE
THIS DESIGN IS THE PROPERTY OF THE FOR MOTOR COMPANY
HYBRIDS PLUS PROVIDES THIS REVERSE ENGINEERED DRAWING TO AID IN PHEV CONVERSIONS
HYBRIDS PLUS GRANTS PERMISSION TO USE THIS DRAWINGTO ANYONE PERFORMING PHEV CONVERSION
UNTIL IF AND WHEN THE FORD MOTOR COMPANY WILL OFFER A PHEV VERSION OF THIS VEHICLE

Rev	Date	Description	By
x1	12/20/05	REVERSE ENGINEERED WIRING	D'de

HYBRIDS-PLUS.COM
Davide Andrea, davide@hybrids-plus.com, 303-413-1500
Title: ESCAPE HYBRID TRACT. BATTERY
Drawing # 200AS000X | C | Sheet 1 of 1

Sources: EAA-PHEV and HybridsPlus.

TABLE 9-2 Hymotion Diagram

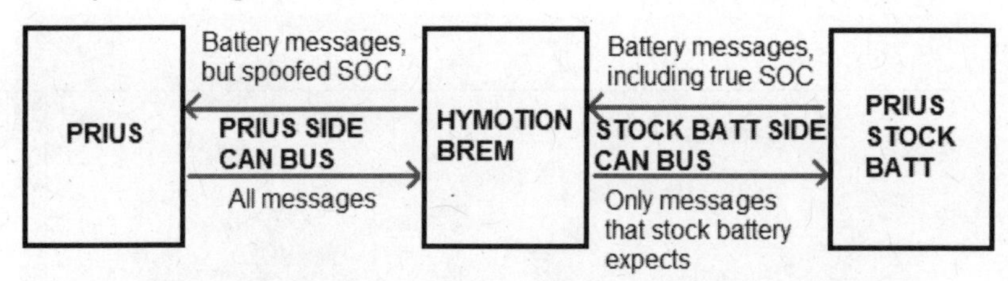

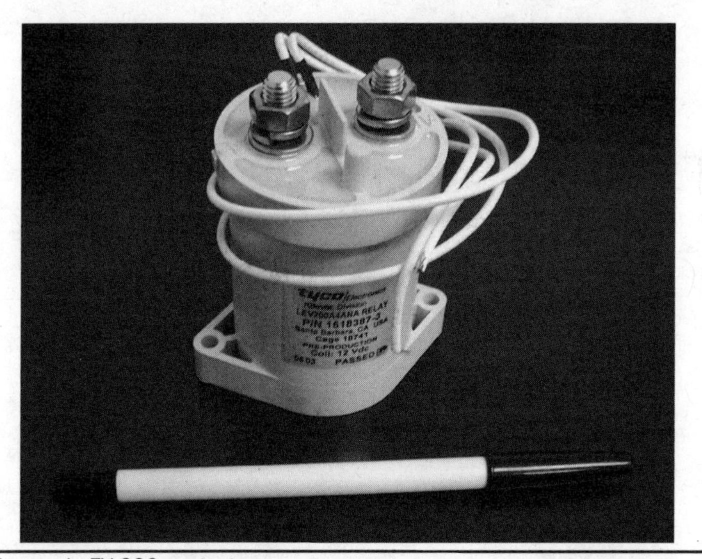

FIGURE 9-8 A sample EV 200 contactor.

FIGURE 9-9 Main contactor box is on the left side of the battery box.

TABLE 9-3 EAA-PHEV Prius Control Board Schematic (Source: EAA-PHEV at http://www.eaa-phev.org/)

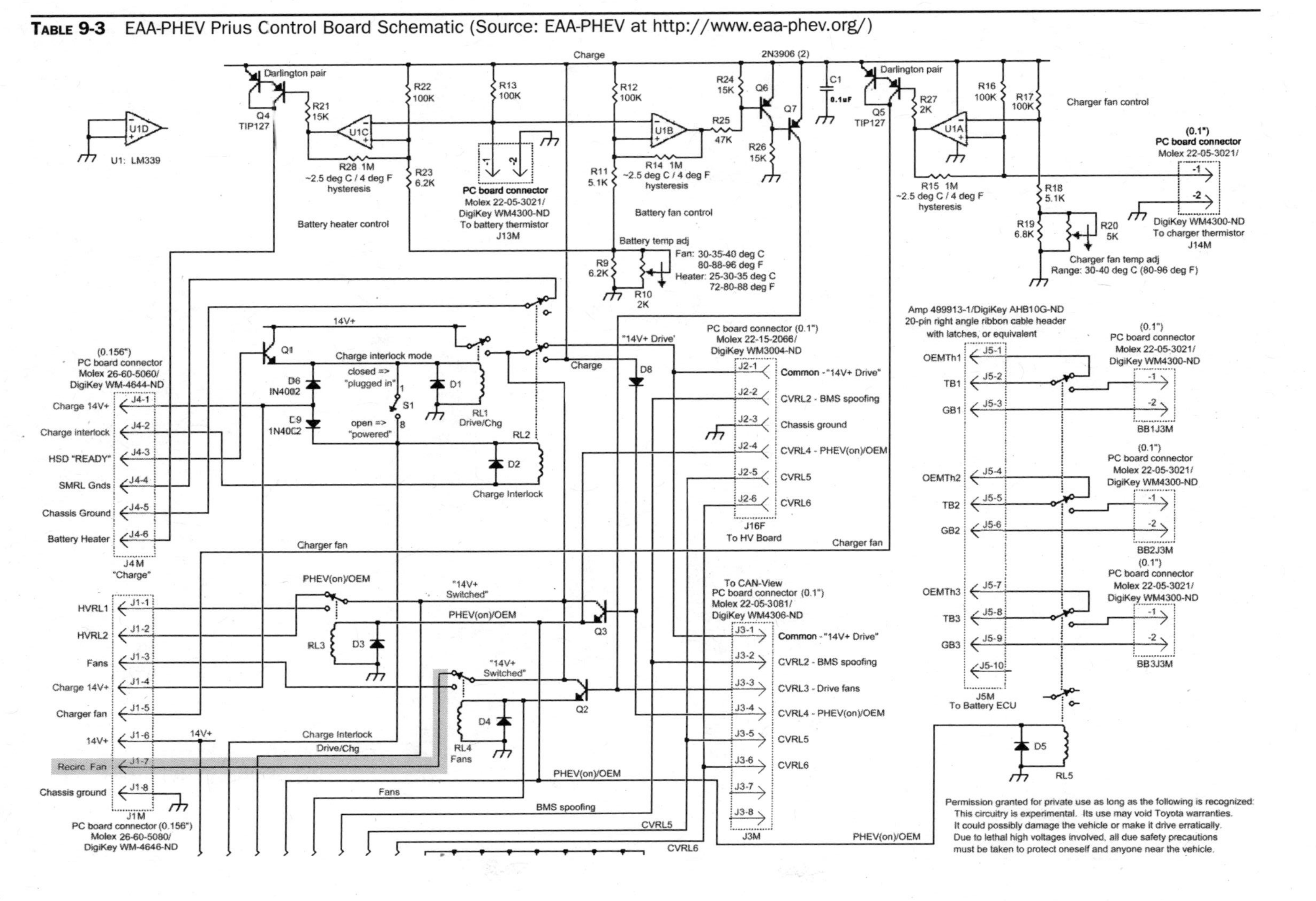

Table 9-4 EAA-PHEV Prius High-Power Schematic (Source: EAA-PHEV at http://www.eaa-phev.org/)

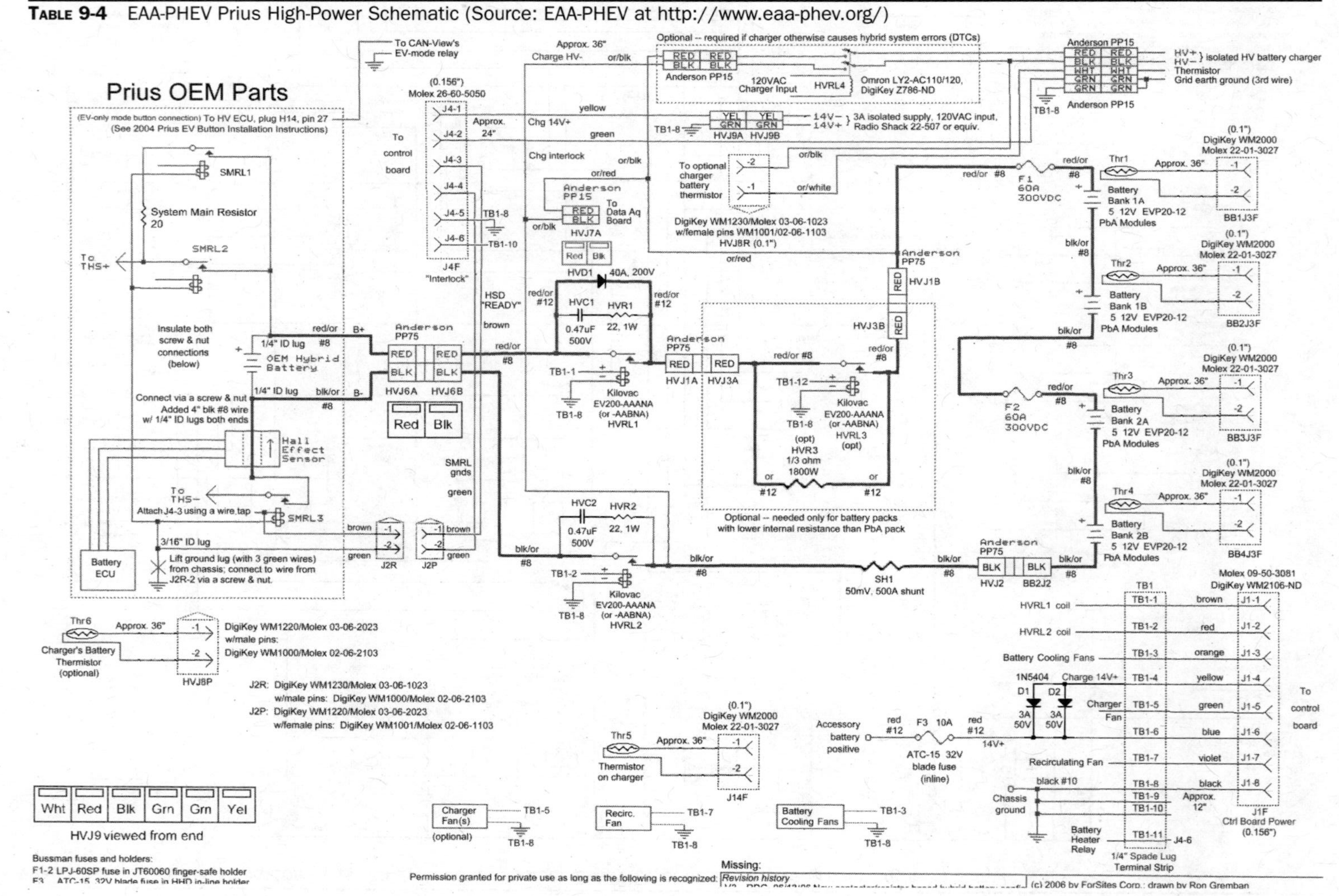

Safety Fuse

The purpose of the safety fuse is to interrupt current flow in the event of an inadvertent short circuit across the battery pack. In other words, you'll blow out one of these before you arc-weld your crescent wrench to the frame and lay waste to your battery pack in the process. Figure 9-10 shows a typical fuse holder.[8]

Safety Interlock

There is an additional switch that some PHEV converters incorporate into their high-current system, usually in the form of a big red knob or button on the dashboard—an emergency safety interlock or "kill switch." When everything else fails, punching this will pull the plug on your battery power.

Low-Voltage, Low-Current Instrumentation System

The instrumentation system includes a key switch, throttle control, and monitoring wiring. The key switch wiring, controlled by an ignition key, routes power from the accessory battery or DC-to-DC converter circuit to everything you need to control when your EV is operating: headlights, interior lights, horn, wipers, fans, radio, and so on. Throttle control wiring is everything connected with the all-important throttle potentiometer function. Monitoring wiring is involved in the remote sensing of current, voltage, temperature, and energy consumed, and in routing this information to dashboard-mounted meters and gauges. Let's take a closer look at these low-voltage components.

Figure 9-10 Typical fuse holder. Source: EAA-PHEV.

Low-Voltage Protection Fuses

All your instrumentation and critical low-voltage components should be protected by 1-ampere fuses (the automotive variety work fine) as shown in Figure 9-10. Whenever 25 cents can save you up to $200, it's a good investment.

Low-Voltage Interlocks

Many EV converters prefer to implement the kill switch referred to earlier in this section on the low-voltage side. This is often easier because there are a number of interlocks there already—seat, battery, impact, and so on. In addition, a low-voltage implementation takes just a simple switch, possibly a relay, and some hookup wire—a few ounces of weight at the most—while a high-current solution takes several pounds of wire plus bending, fitting, and so on.

Wiring It All Together

Four things are important here—wire and connector gauge, connections, routing, and grounding. We'll cover them in sequence.

However, to help out, Tables 9-1 through 9-4 show wiring diagrams for the Ford Escape Hybrid and the Toyota Prius, the two most popular PHEV conversion vehicles on the market today.

Wire and Connectors

This may be one of the last things you think about, but it's by no means the least important. While your choices of wire size and connector type on the instrumentation side are not as important as the connections you make with them, all of these are important on the power side.

Minimal resistance means that how the connectors are attached to the wire cable ends is equally important for the overall result. Crimp the connectors onto the cable ends using the proper crimping tool (ask your local electrical supply house or cable provider) or have someone do it for you. If you are getting 20 miles per charge and your neighbor is getting 60 miles per charge with the identical setup, and you've checked for the obvious mechanical reasons—motor, controller, and battery—chances are that the problem is in your wiring. Treat each crimp with loving attention and craftsmanship, as if each were your last earthly act, and you will be in heaven when it comes to your EV's performance.[9]

Connectors

AC-to-DC Connector—Example from Ford Escape Hybrid

Figure 9-11 shows the AC-to-DC converter.[10] The AC-to-DC converter connector has two circuits, with the following names and functions.

This connector is on one end of a cable. The other end of the cable (C1468) is capped, under the hood on the right, in front of the two coolant tanks, fastened to

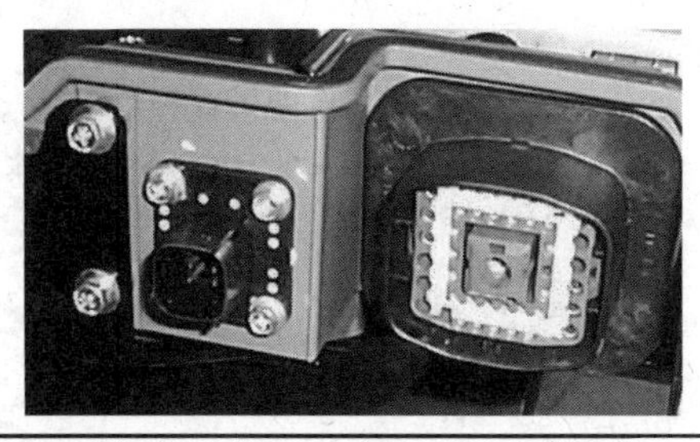

FIGURE 9-11 AC-to-DC connector (left) and control connector (right).

its own harness. It is used only with the engine block heater option, which includes a small 115 V AC–to–24 V DC converter. Ford's part numbers for the block heater option are 4M6Z-6B018-AA WIRE ASY, F5RZ-6A051-B HEATER ASY, and 5M6Z-10B689-AA CHARGER ASY. When the engine block heater is plugged into the 115 V AC, a "Y" splitter sends some power to the AC-to-DC converter and then to the traction battery. It might charge only when the HV battery reaches a low state of charge, or it may simply be a battery warmer, because under normal battery conditions and room temperatures, zero power is sent to the traction battery pack. The engine block heater is rated at 115 V AC and 400 watts. The AC-to-DC converter output is 24 V DC under no load, and it draws 75 watts when it is plugged into a cold battery. Upon initial testing, a 7°C HV battery was warmed to 32°C in about three hours.[11]

Table 9-5 shows the AC-to-DC converter connector and its two circuits, with their respective names and functions.[12]

TABLE 9-5 AC-to-DC Converter and the Names and Functions

ID (hex)	Period [ms]	No of data bytes	Byte 0	Byte 1	Byte 2	Byte 3	Byte 4	Byte 5	Byte 6	Byte 7
300h	10	5	Current		Voltage	Flags	00h			
310h	100	7	Constant	Constant	Constant	Constant	Temp-erature	Charge Limit	Discharge Limit	
320h	100	5	(DTCs?)	(DTCs?)	Flags	SOC				

Source: EAA-PHEV.

Control Connector C4227A—Ford Escape Hybrid PHEV

The control connector has 40 positions, but only 24 circuits (see Figure 9-12). To disconnect it, turn the bolt, which draws the connector out. To remove the bulkhead male from the battery, remove the black shroud, squeeze the two gray snaps left and right, and pull into the battery body. To remove a pin from either mate, look on the mating surface, find the white rectangular plastic retainer, use a small flat screwdriver to lift the little snaps, and lift the retainer. On the wire side, pull on the wire for that pin; on the pin side, use the small screwdriver to release the gray plastic snap holding the pin. Pull the wire and the pin out.[13]

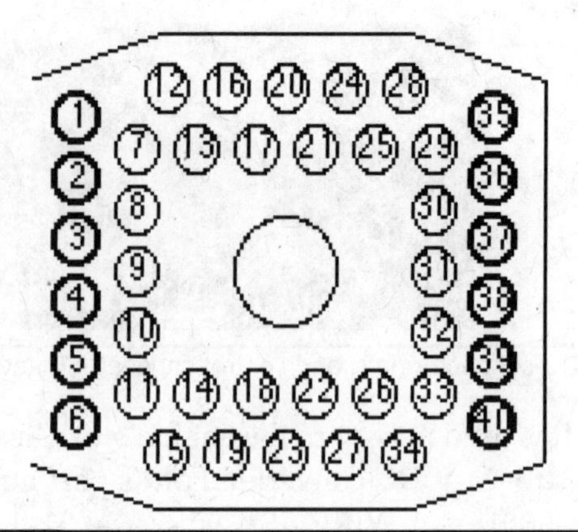

Figure 9-12 Pin-out, looking into cable (left) and looking into battery (right).

Immediate Shutdown

With these two lines, the battery tells the transaxle control module that all is OK. However, whenever there's 12 V on the start/run and all is OK, the battery sends 12 V to both immediate shutdown lines. The load in the transaxle control module on each line is 1.2 kohms. If *both* lines are open, the transaxle control module shows a fault (if only one line is open, then all is OK). Figure 9-13 shows the immediate shutdown circuit for the Ford Escape Hybrid.

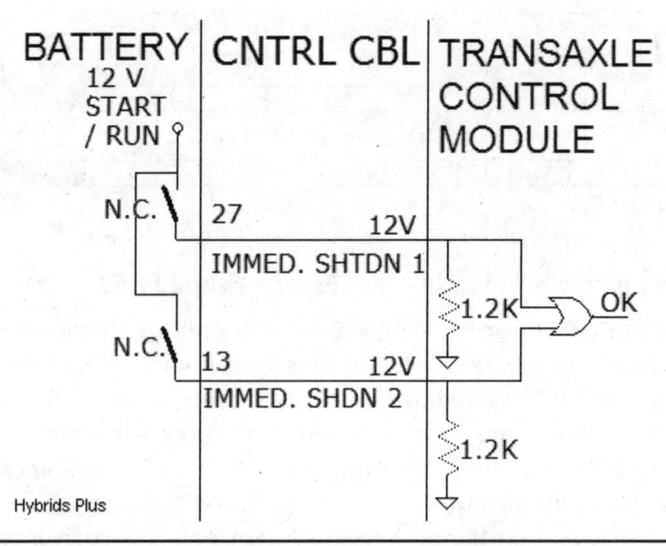

Figure 9-13 Immediate shutdown circuit for the Ford Escape Hybrid.

HV Connector C4227C

Figure 9-14 is a picture of the Ford Escape battery pack power and data connector from Yazaki. The specifications for the connectors are:

- Male (on battery) P/N 7325-6498-02 or 7325-6499-02
- Female (on cable) P/N 7325-6490-51

Figure 9-15 shows the spec sheet directly from Yazaki. It shows the connector and the plastic protected connection between the metal casings.

Figure 9-16 shows the Ford Escape HV interlock circuit. The circuit goes from the battery through the transaxle control module and back to the battery. If either wire is opened, shorted to +12 V, or grounded, both the battery and the transaxle control module detect a fault.

Contactors

Contactors Assembly

This assembly includes two high-power contactors. Figure 9-17 shows the Ford Escape safety disconnect socket and plug. Figure 9-18 shows the battery safety disconnect socket, with the shunt on the other side and a low-current precharge relay. Figure 9-19 shows the contactor assembly and the precharge resistor (dangling, in the picture).

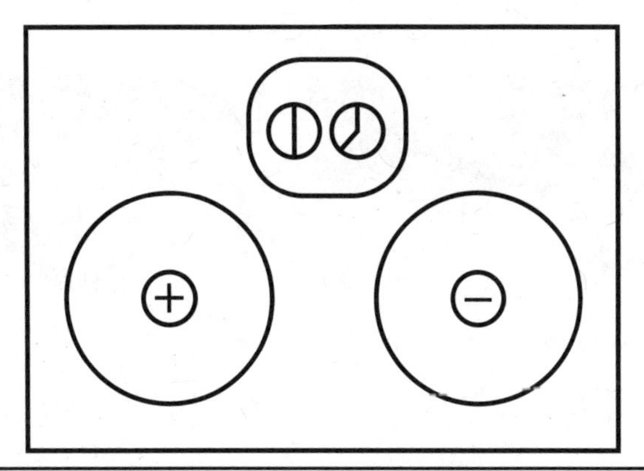

FIGURE 9-14 Ford Escape data connectors. Source: http://www.eaa-phev.org/wiki/Image:Batterypack-power-conn.jpg

All Systems Go!™

2P+2p DC Power Connector

Description:

High Voltage, sealed, electromagnetically shielded connector for Battery/Inverter applications on electric drive systems

Features:

- Electromagnetically shielded connector
- Contact reliability ensured by using multi-contact spring terminal
- Last to mate/first to break safety interlock circuits
- Bolt on module side connector

Benefits:

- No separate drain wires required for grounding EMI induced current
- Lower insertion force due to use of lever
- Safety interlock circuits assure customer safety
- Module side connector easily integrates into Motor or Inverter case

Specifications:

Part Number	7325-6498-02 7325-6499-02	7325-6490-51
Type	Unit (Male) Side	W/H (Female) Side
Pole	2P + 2p	2P + 2p
Voltage Capacity	DC600V	DC600V
Current Capacity	120A	120A
Applicable Wire Size	N/A	EEXBS 20mm² EEXBS 15mm²
Shield Effectiveness	40dB	40dB

Status: *In Production*

For more information, contact info@us.yazaki.com

YAZAKI NORTH AMERICA, INC. • 6801 HAGGERTY ROAD • CANTON, MICHIGAN 48187 • www.yazaki-na.com

Figure 9-15 Spec sheet from Yazaki. Source: Yazaki North America Inc.

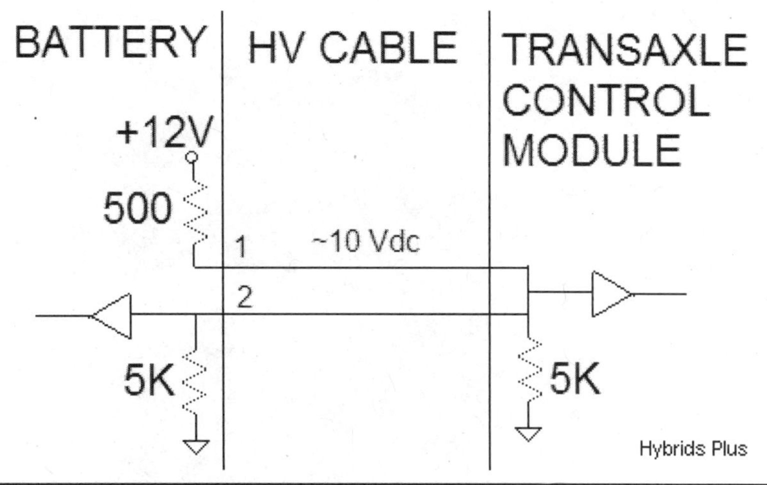

FIGURE 9-16 The Ford Escape HV interlock circuit. Source: EAA-PHEV.

FIGURE 9-17 One side of the safety disconnect for the Ford Escape battery.

Figure 9-18 The other side of the Ford Escape safety disconnect socket.

Figure 9-19 Contactor assembly.

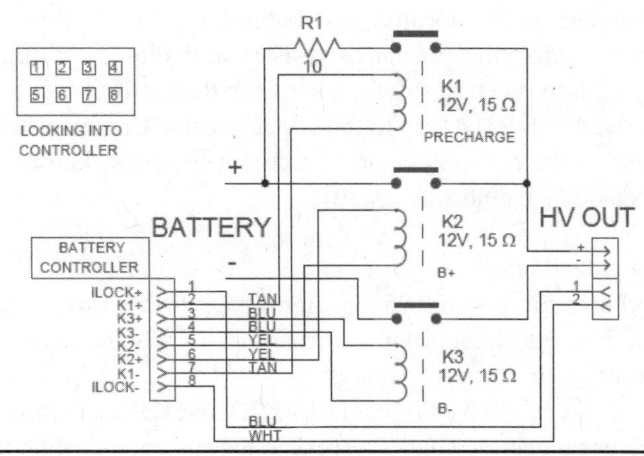

Figure 9-20 Ford Escape Hybrid—traction battery contactors circuit. Source: EAA-PHEV.

The traction battery uses three contactors (high-power relays), as shown in Figure 9-20, to connect the battery voltage to the HV output. Figure 9-21 shows the sequence of contactors for the Ford Escape Hybrid traction battery.

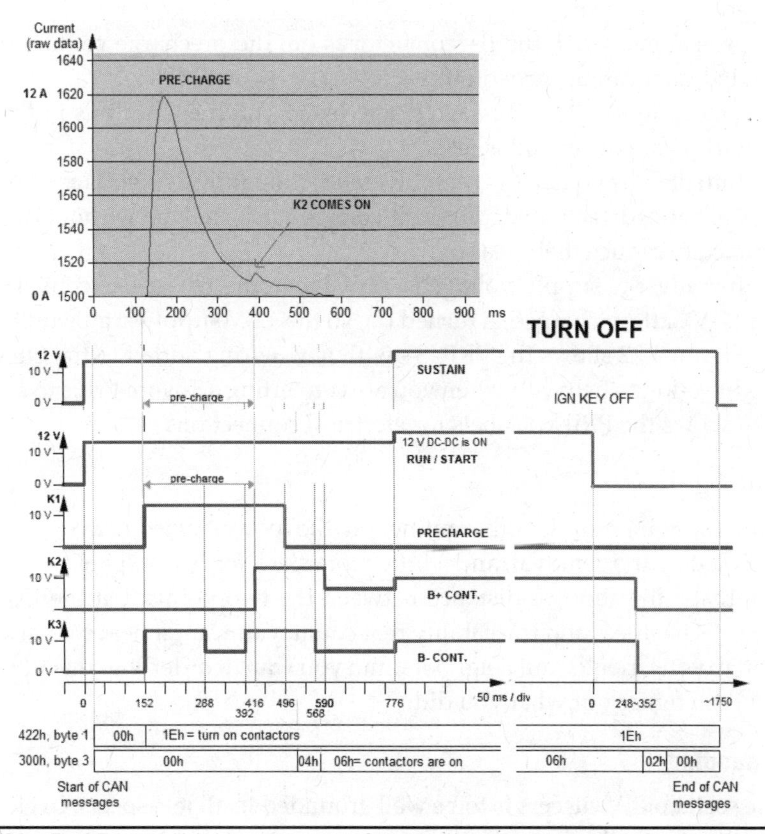

Figure 9-21 Ford Escape Hybrid—sequence of traction battery contactors. Source: EAA-PHEV.

The contactors are located just behind the HV output connector, and the enclosure includes one precharge resistor and filter capacitors. The purpose of contactor K1 is to precharge the capacitors in the motor driver slowly. Contactor K2 connects the B+ of the battery to the HV connector. Contactor K3 connects the B– of the battery to the HV connector. The circuit includes the interlock that detects a disconnected HV connector.

Sequence of Contactors

The CAN bus activity starts 25 ms after the ignition is turned on.

At 150 ms after the ignition is turned on, a CAN message (422h, byte 1 = 1Eh) tells the battery to turn on its relays.

Then the precharge and B– contactors (K1 and K3) are turned on to precharge the motor driver capacitors. There's a spike in the current reflecting the inrush. The time constant is measured to be about 35 ms. Given that the precharge resistor is 10 ohms, we derive that the capacitors in the inverters are 3.5 mF (that's 3.5 millifarads).

At 416 ms, the B+ contactor (K2) is turned on to apply the full battery voltage to the motor driver. There's a small step in the current, as the precharge resistor is no longer in the circuit.

At 496 ms, when the B+ contactor is on, the precharge contactor is no longer needed, so it can be turned off.

The battery puts a message (300h, byte 3) on the CAN bus indicating that the contactors are on and all is OK.

Initially, the contactors are powered by the full 12-V voltage. After a bit, since they are already actuated, their coil voltage can be halved without dropping off the contact, to reduce their heating.

Initially, the supply voltage is 12 V. Later, the DC-to-DC converter that keeps the 12-V battery charged is turned on, so the 12-V supply jumps up to 14 V.

Figure 9-22 shows the PRIUS+ with new wiring added to the OEM battery fan modification to help you when you are converting a Toyota Prius to a PHEV. Figure 9-23 shows the PRIUS+ labels for electrical connections.

Routing

Aim for minimum-length routing on the power side. Leave a little slack for installation and removal, and a little more slack for heat expansion, then go for the line that's the shortest distance between the two points. On the instrumentation side, it's neatness and traceability that count: you want it neat so that you can show it off to your friends and neighbors, and you want it orderly so that you (or someone else) can figure out what you did.

Grounding

The secret of EV success is to be well grounded in all its aspects. In electrical terms, "well grounded" means three things:

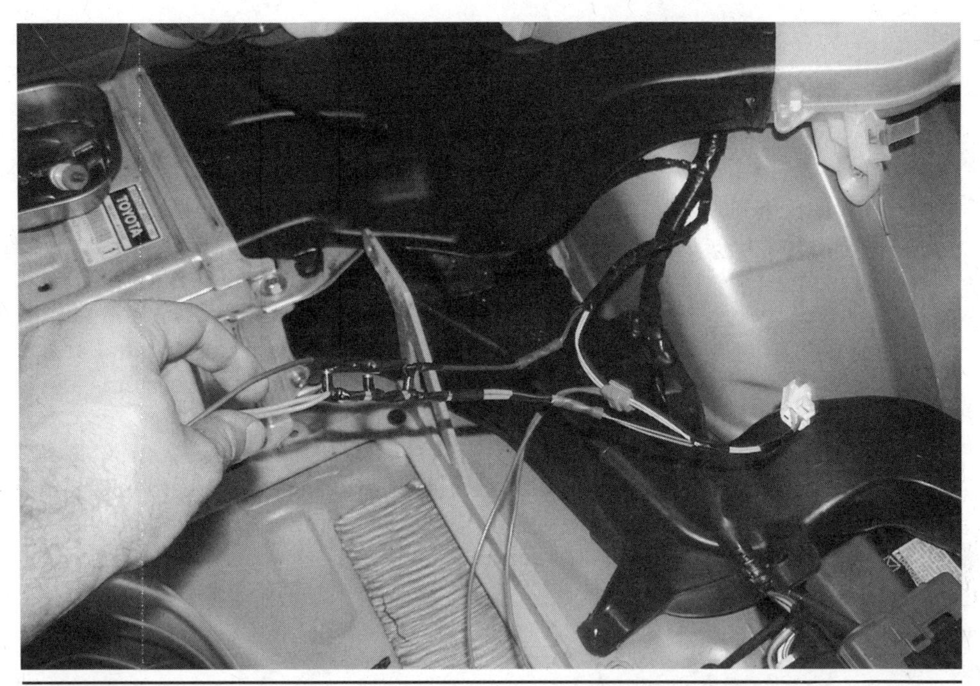

FIGURE 9-22 New wiring added to the OEM battery fan modification. Source: www.eaa-phev.org.

FIGURE 9-23 PRIUS+ labels for electrical connections.

- *Floating propulsion system ground.* No part of the propulsion system (batteries, controller, or anything else) should be connected to any part of the vehicle frame. This minimizes the possibility of your being shocked when you touch a battery terminal and the body or frame, and of a short circuit occurring if any part of the wiring becomes frayed and touches the frame or the body.
- *Accessory 12-volt system grounded to frame.* The 12-volt accessory system in most EV conversions is grounded to the frame, just like the electrical system of the internal combustion vehicle chassis it utilizes. The body and frame are not connected to the propulsion system, but you can and should use them as the ground point for the 12-volt accessory system, just as the original vehicle chassis manufacturer did.
- *The electric circuit for the HV interlock.* The circuit goes from the battery through the transaxle control module, and back to the battery. If either wire is opened, shorted to +12 V, or grounded, both the battery and the transaxle control module detect a fault.[14]

Conclusion

While this chapter was designed to show people the main points to remember when converting a car, always read the material on the EAA-PHEV site. There is much more detailed information on this site for people who want to convert their vehicle.

CAN bus language is essential and needs to be programmed exactly correctly, so make sure that you have a manual and that you have really read all the information on the EAA-PHEV site.

If you get a conversion kit from a reputable company, such as Elithion or PRIUS+, the company will be there to help you with your conversion.

CHAPTER 10

Plug-In Hybrid Electric Vehicle Conversion

Now, whether you have read through the whole book or not, this is the chapter that tells you how to do your PHEV conversion and make it run. You've learned about choosing the right drive system and how to design your PHEV. Assuming you have read the preceding chapters, you've also learned about motor, controller, battery, charger, and system wiring recommendations. Now, it's time to put all these things together. A carefully planned and executed conversion process can save you time and money, and produce an efficient vehicle that's a pleasure to drive and own after it's completed.

This chapter goes through the conversion process step by step with the assistance of a few conversion specialists. It also introduces the type of chassis to choose for your PHEV conversion today—the Toyota Prius or the Ford Escape Hybrid.

You'll discover that after the simple act of going through the conversion process, your efforts and results will perform even better and you are now an expert.

Conversion Overview

What do you do to get a fresh point of view when you're constructing something mechanical, even if it's a plug-in hybrid electric vehicle?

You go to a mechanic!

What do you do first? Start with a hybrid car (to make it easy here!). As we should all know, the Toyota Prius is the most popular hybrid on the market. Conversion specialists can make an excellent platform for PHEVs because they have a drive system or CAN bus that is easy to communicate with to make a PHEV.

Keep in mind that, while we are talking about a step-by-step Toyota Prius and Ford Escape PHEV conversion here, the principles will apply equally well to any sort of PHEV conversion you do.

Figures 10-1 and 10-2 show you the entire process at a glance. Let's get started.

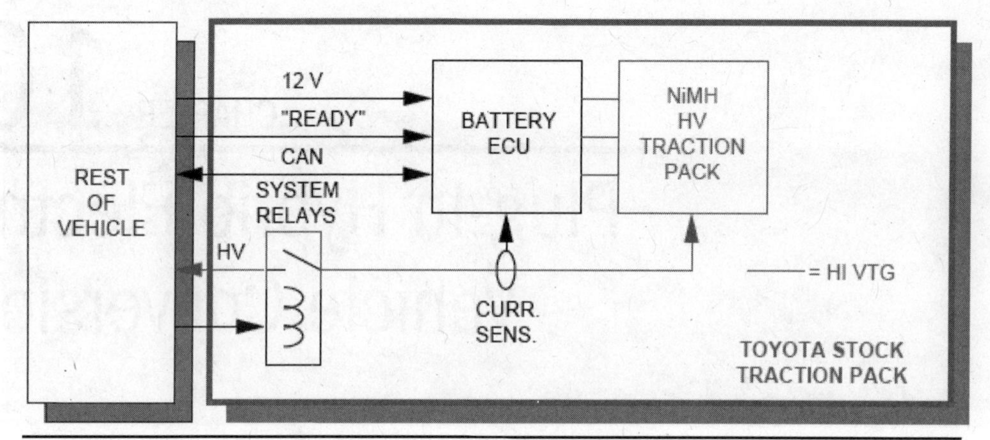

FIGURE 10-1 Block diagram of a Toyota Prius hybrid electric vehicle. Source: EAA-PHEV.

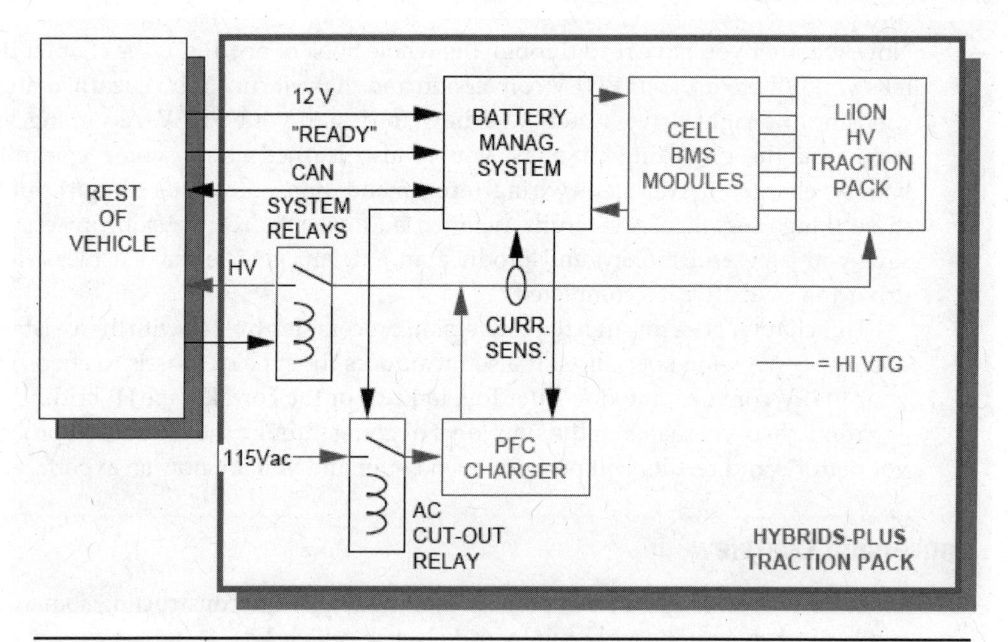

FIGURE 10-2 Block diagram of a Ford Escape plug-in hybrid electric vehicle from Hybrids Plus. Source: EAA-PHEV.

What's the Objective?

The objective here is to get you into a working PHEV of your own with minimum fuss, converting it from an internal combustion engine vehicle chassis. For those who are into building from-the-ground-up and kit-car projects, there are other books you can read, and the techniques discussed here can be adapted. The actual process for your PHEV conversion is straightforward:

- Before conversion (planning)—who, where, what, when
- Conversion (doing)—chassis, mechanical, electrical, and battery
- After conversion (checking)—testing and finishing

Before Conversion

A little effort expended before conversion,[1] in the who-where-what-when planning stage, can pay large dividends later because you'll have thought out what you need beforehand, and you won't have to go running around at the last minute. We'll look at the individual areas.

Conversion

Conversion planning can pay even greater dividends. There are four parts to the conversion, or doing stage, and each is further subdivided:

- *Chassis*. Purchase, preparation, and removal of internal combustion engine parts
- *Mechanical*. Motor mount fabrication, motor installation, battery mounts, and fabrication and installation of other mechanical parts
- *Electrical*. High-current, low-voltage, and charging system components and wiring
- *Battery*. Purchase and installation of batteries

A simple way of looking at the procedure is: buy and clean up the chassis, remove all the internal combustion engine parts, make or buy the parts needed to mount the motor and batteries, mount and wire the electrical parts, then buy and install the batteries. Let's look at the individual areas.

PRIUS+ History[2]

Ron Gremban was the technical lead for CalCars's 2004 PRIUS+ modification of my 2004 Prius into a plug-in hybrid (PHEV), and he remains CalCars's technical lead. Felix Kramer, CalCars's founder, enlisted me in mid-2004, shortly after I first contacted him. The project had a lot of Internet collaboration as well as physical volunteer help, largely from Electric Automobile Association (EAA) members (thanks to everyone). You can view much of the history of the project at http://autos.groups.yahoo.com/group/priusplus/messages, especially the summary at http://autos.groups.yahoo.com/group/priusplus/message/421. (This group is currently closed to new members, but postings are viewable by nonmembers.)

Our first attempt to add more battery capacity to Toyota's hybrid system (not unexpectedly) didn't work. We discovered that Toyota's battery management computer (battery ECU) uses ampere-hour integration as its major way of determining the battery's state of charge (SOC). Therefore, just as with the OEM battery, it would pop the vehicle out of EV-only mode after a mile or so because of a very low SOC indication (in spite of the new battery's much larger capacity and high actual SOC).

Greg Hanssen offered EnergyCS's help at low cost. It had a CAN bus controller that it turned into a battery management computer to replace Toyota's battery ECU by emulating that OEM computer's CAN messages to the rest of the hybrid system. This system, though proprietary, got our PRIUS+ working quickly; first smoke-tested in late September, it was working well by early November 2004.

Greg and others at EnergyCS then got excited, built their own Prius conversion using Valence lithium ion batteries, and—with CleanTech, a company to which Felix introduced them—started EDrive Systems to commercialize Prius PHEV conversions. This is an exciting development, which may provide, for most people, the first opportunity to own a PHEV.

However, there is also a community of experimenters, schools, and others who are interested in doing such modifications themselves, and the EnergyCS controller is not available for this purpose. Fortunately, Dan Kroushl discovered and shared the information that a higher-voltage battery is capable of fooling Toyota's battery ECU into giving a higher SOC reading. I started experimenting with higher voltages of lead-acid batteries and came up with an algorithm for fooling the battery ECU in an organized way. I tried it first with lead-acid batteries and only with manual control, but I expected it to work more generally. It will be described in detail later.

I have now completed three schematic diagrams for a PRIUS+ conversion using this technique, and I am working on software to automate the control as well as to display and record relevant vehicle information. This is intended to run first on a laptop computer, then on a PocketPC PDA that can be semipermanently mounted (and left running) in the car.

Battery Pack

Ron specified a BB battery lead-acid battery pack (18 EVP20-12B1 modules from ElectricRider.com) for our first conversion attempt, in order to bring down our learning curve with inexpensive, fairly indestructible batteries. This turned out to be an excellent decision; although this pack required replacement after 200 cycles and just under a year, it has sustained our PRIUS+ through more than a year of work on finding a more advanced battery pack. The BB electric bicycle batteries, though very marginal for this application, are actually quite remarkable among lead-acid batteries. When fully charged, these 20-Ah (12-Ah @ ½-hour rate) batteries have a lower internal resistance (~250 milliohms) than the OEM battery (300–350 milliohms) and are capable of supplying the Prius's maximum draw of 200 A—a 10°C or 17°C rate, depending on which capacity you calculate it against. At low SOC, however, the internal resistance is >500 milliohms, too high for full hybrid functionality. My PRIUS+'s performance has been documented in CalCars's

PRIUS+ Fact Sheet. In short, the BB batteries weigh about 260 pounds and provide for approximately double the normal Prius's gasoline mileage for around 20 miles of mixed-mode (town and highway) driving on each charge.

Next, Ron planned on installing a nickel–metal hydride (NiMH) pack, largely to prove that the chemistry that is already in all of today's mass-produced hybrids is also capable of powering electrically and economically effective PHEVs, thereby removing one more auto company excuse for not building them. As soon as we have the people resources, we will move on to work with one or more of the newer high-performance lithium ion suppliers, although we expect these batteries to require additional engineering work to ensure proper cell-by-cell charge and discharge control as well as to engineer out any likelihood of thermal runaway problems, to which this chemistry is prone.

Ron spent the first nine months of 2005 looking for NiMH and/or lithium ion batteries that fit the PRIUS+ requirements. In the process, I learned a lot about the characteristics of these advanced battery chemistries. For example, special problems are encountered when NiMH cells or strings are put in parallel. This should be done only with full strings, and with special control electronics such as those designed for the Electro Energy pack (discussed later in this chapter). Wayne Brown and Dan Kroushl are using many parallel strings of NiMH sub-C cells in their enhanced (but not PHEV) Priuses. These small (3–3.6-Ah) cells have good high-power capabilities, and the parallel strings work well in their application where the cells do not get fully charged.

Ron figured that D cells, also quite available, would require fewer parallel strings so that the electronics might not be overwhelming. However, although many of them looked good on paper, all the D cells Ron tested had at least double the expected internal resistance. Fortunately, Electro Energy came along with its specialized and capable NiMH batteries.

Ron also looked at many lithium ion possibilities. The least expensive option is currently 18650 cells, produced in huge volumes for laptop computers. However, these, like the sub-C NiMH cells, are very small (2–2.8 Ah). Also, they are typically designed for 2- to 5-hour discharge rates and lack high-power capabilities. You pay more for high power, low susceptibility to thermal runaway, and long cycle life—and you still have to use many cells in parallel to reach the necessary capacity. A typical PRIUS+ pack would use at least 1,200 cells, providing plenty of failure points.[3]

CalCars and Electro Energy, Inc.

CalCars and Electro Energy, Inc., in Danbury, Connecticut, are now in the late stages of a joint project to adapt their unique form of NiMH packs to the PRIUS+. As part of the project, I have designed control circuitry that is specific to the characteristics of this battery and to Toyota's battery management system. When complete, the pack should weigh about the same as the BB battery pack but last for 40 to 50 miles of mixed-mode driving. The internal resistance figures that we have seen so far are in the 200-milliohm range. It is possible that Electro Energy will elect to sell this pack to experimenters, so stay tuned.

Lead-Acid Batteries

Several EAA members have recommended Hawker lead-acid batteries, which don't appear to exist anymore under that moniker. However, I believe I just tracked down their descendants: Odyssey batteries by EnerSys Inc. This firm has a model PC625 that may be less marginal than the EVP20 and therefore useful for experimenters' PRIUS+ conversions. I don't yet know the price or the availability. Although these batteries are rated at 16 Ah, they have the same rated ½-hour capacity as the EVP20, are 1 pound lighter, and have a claimed 1,800-A short-circuit current at full charge, requiring an internal resistance of less than 7 milliohms (126 milliohms for a full pack of 18). Also, they claim an 80 percent DOD cycle life of 500 and the ability to withstand long periods of discharge with little damage.

Several PRIUS+ battery pack specifications are provided at www.eaa-phev.org/wiki/PriusPlus_History#Appendix_A:_Battery_Specifications.

High-Voltage Circuit

The high-voltage circuits[4] consist largely of the PHEV battery pack, some contactors, a circuit to emulate the OEM battery's taps, a charger, and miscellaneous battery support devices. The OEM battery of a hybrid electric vehicle is not in the circuit and can be removed to reduce the vehicle's weight by nearly 100 pounds. However, the circuitry is designed to make reconnecting the OEM battery easy, and it is recommended that it be kept in the vehicle until the reliability of your conversion has been proven to your satisfaction.

For safety and ease of installation/removal, the battery pack has been divided into two banks, separately fused and used in series, and connected together only during operation. With the current BB battery pack, this corresponds to two physical rows of battery modules. We hope that this arrangement will work for many other batteries as well, though the Electro Energy pack is specialized and requires very different high-voltage and control electronics.

Colors of High-Voltage Wiring

The color of high-voltage wiring is shown as red/orange or black/orange, depending on the polarity. This is because hybrid vehicles have standardized on orange to indicate high-voltage circuitry to service and emergency personnel, but it is desirable to indicate polarity as well. Though orange welding cable is not commonly available, the ends can be wrapped in orange tape or have orange tie wraps added. Number 6 welding cable should be sufficient for the high-power cables. Anderson connectors have been specified for all high-voltage connections, as both ends can be hot when disconnected. A Prius's HV battery is electrically isolated from the vehicle's chassis.

TABLE 10-1 High-Power Schematic 2d

Reference	QTY	Description	Vendor	Vendor #	DigiKey #	Cost	Price	Comments
R1–R14	14	1K Ohm 0.5W 1% Metal Film Resistor	Phoenix Passive Components	5033ED1K000F12AF5	PPC1.00KXCT-ND	$2.72	$0.19	
R16, R24–27	5	100K Ohm 0.25W 5% Carbon Film Resistor	Panasonic	ERD-S2TJ104V	P100KBACT-ND	$0.34	$0.07	
R17	1	10K Ohm 0.25W 5% Carbon Film Resistor	Panasonic	ERD-S2TJ103V	P10KBACT-ND	$0.07	$0.07	
R18	1	49.9K Ohm 0.5W 1% Metal Film Resistor	Phoenix Passive Components	5033ED49K90F12AF5	PPC49.9KXCT-ND	$0.19	$0.19	
R19, R30	2	200K Ohm 0.25W 5% Carbon Film Resistor	Panasonic	ERD-S2TJ204V	P200KBACT-ND	$0.13	$0.07	
R20	1	160K Ohm 0.25W 5% Carbon Film Resistor	Panasonic	ERD-S2TJ164V	P160KBACT-ND	$0.07	$0.07	
R21, R23	2	47K Ohm 0.25W 5% Carbon Film Resistor	Panasonic	ERD-S2TJ473V	P47KBACT-ND	$0.13	$0.07	
R29	1	56K Ohm 0.25W 5% Carbon Film Resistor	Panasonic	ERD-S2TJ563V	P56KBACT-ND	$0.07	$0.07	

(continued)

TABLE 10-1 High-Power Schematic 2d *(continued)*

Reference	QTY	Description	Vendor	Vendor #	DigiKey #	Cost	Price	Comments
R33	1	470 Ohm 0.25W 5% Carbon Film Resistor	Panasonic	ERD-S2TJ471V	P470BACT-ND	$0.07	$0.07	
R35	1	470K Ohm 0.25W 5% Carbon Film Resistor	Panasonic	ERD-S2TJ474V	ERD-S2TJ474V	$0.07	$0.07	
R31	1	200K Ohm 0.5W 1% Metal Film Resistor	Phoenix Passive Components	5033ED200K0F12AF5	PPC200KXCT-ND	$0.19	$0.19	
R36	1	1.0 Meg Ohm 0.5W 1% Metal Film Resistor	Phoenix Passive Components	5033ED1M000F12AF5	PPC1.00MXCT-ND	$0.19	$0.19	
C1–C15	15	0.1 uF Metallized Polypropylene Film Cap	Epcos	B32621A5104J	495-1375-ND	$8.10	$0.54	Radial leads, 10 mm spacing, Very good self-healing properties
C16	1	0.01 uF Metallized Polypropylene Film Cap	Epcos	B32621A3103J	495-1283-ND	$0.37	$0.37	Radial leads, 10 mm spacing
R15, R28	2	25K Ohm 1 turn trimpot–top adjust	MuRata	PVC6M253C01B00	490-2822-ND	$2.10	$1.05	
R22, R34	2	50K Ohm 1 turn trimpot–top adjust	MuRata	PVC6M503C01B00	490-2827-ND	$2.10	$1.05	

Table 10-1 High-Power Schematic 2d *(continued)*

RL1	1	SPST-NO DIP Reed Relay– 12V with diode	Hamlin	HE721A1210	HE103-ND	$1.79	$1.79	
RL2–RL5	4	4PDT Power Relay 3A 12VDC	Omron	MY4-DC12(S)	Z186-ND	$25.92	$6.48	Elliptical leads
RS2–RS5	4	Relay Socket for RL2–RL5	Omron	PY14-02	Z813-ND	$7.20	$1.80	PCB Mounting
D2	1	Silicon Rectifier – 100V, 1 A	Diodes, Inc.	1N4002-T	1N4002DICT-ND	$0.26	$0.26	DO-41 Package
D4	1	Silicon Rectifier –200V, 1 A	Diodes, Inc.	1N4003-T	1N4003DICT-ND	$0.26	$0.26	DO-41 Package
D1, D3, D5–D9	7	High Speed Diode – 1N4148	Philips	1N4148 T/R	568-1360-1-ND	$0.35	$0.05	DO-35 Package
D10	1	91 Volt Zener Diode – 2 W, 5%	MicroSemi	2EZ91D5DO41	2EZ91D5DO41MSCT-ND	$1.42	$1.42	DO-41 Package
D11	1	120 Volt Zener Diode – 2 W, 5%	MicroSemi	2EZ120D5DO41	2EZ120D5DO41MSCT-ND	$1.42	$1.42	DO-41 (ackage
Q1, Q4, Q5	3	NPN Darlington Transistor, 5 A, 100 V	Fairchild	TIP122	TIP122FS-ND	$1.80	$0.60	TO-220 Package
Q2, Q6, Q7	3	NPN Transistor, 500 mA, 300 V	Fairchild	MPSA42	MPSA42FS-ND	$1.08	$0.36	TO-92 Package
Q8	1	PNP Transistor, 500 mA, 300 V	Fairchild	MPSA92	MPSA92-ND	$0.36	$0.36	TO-92 Package
J8M	1	16 pin right angle ribbon cable header with latches	Amp	499913-3	AHB16G-ND	$2.20	$2.20	

(continued)

TABLE 10-1 High-Power Schematic 2d *(continued)*

J7M	1	20 pin right angle ribbon cable header with latches	Amp	499913-4	AHB20G-ND	$5.18	$5.18	
J2M	1	6 pin Connector to Control Board	Molex	22-05-3061	WM4804-ND	$1.23	$1.23	
J13M	1	5 Position Header Connector .156" right angle	Molex	26-60-5050	WM4643-ND	$0.64	$0.64	
PS1	1	Isolation Power Supply 12 VDC to ± 15 VDC	V-Infinity	PTK10-Q24-D15	102-1233-ND	$39.24	$39.24	1 in. x 2 in. x 0.4 in.
					Total cost	**$107.26**		

Electric Vehicle Charger

The charger to use will depend on the specifics of the battery pack. Choose one with the proper charge control and shutoff scheme for your specific pack. A 4-kW, 3-hour BRUSA NLG5 charger from MetricMind.com was used for our first conversion. A 1- to 1.5-kW capacity is usually sufficient and advantageous, as it is no problem if it takes all night to charge a Prius, and this capacity can be supplied from an ordinary U.S. 120-V AC, 15-A circuit. A charger with an electrically isolated output is highly recommended. The charger's third-wire ground should be connected to the vehicle's chassis, and it is recommended that the vehicle's home charging circuit be GFI protected. Delta-Q is supplying chargers to EDrive Systems.

The charger also needs a 5-A, 14–14.5-V power supply that is capable of powering the control electronics and fans, activating HVRL1, and supplying a trickle charge to the Prius's accessory battery as well. The latter is needed because it has been found to be too easy to leave a dome light or other accessory on during charging, and it is inadvisable to deeply discharge the vehicle's accessory battery. A 5-A, 12-V DC supply is available for $30 from www.amondotech.com. I suspect it can be opened and modified (e.g., by adding three signal diodes in series with its zener diode) to put out 14 V DC; I did this to a similar unit from Electronics Plus.

When the vehicle is off and not charging, the HVRL1 is open, the PHEV battery is not connected to the hybrid system, and only half the 216 V nominal voltage appears anywhere—still high and dangerous, but not nearly as much as the full pack voltage.

Fuses

Fuses F1 and F2, and F3 and F4, work in parallel to protect each battery bank from electrical system shorts. The parallel fuses were specified because fuses with high-voltage DC ratings are rare and their fuse blocks are usually large, with no insulation to protect from shorts and touching. The Bussman LPJ-60SP fuses, in contrast, are rated for 300 V DC and have small, insulated fuse blocks from which the fuses can be removed without touching live circuitry—but these fuse holders are available only for fuses up to 60 A, about half of the PRIUS+'s requirement, given the fuse's particular speed of operation. In an installation using them, it is therefore necessary to ensure that the wiring resistance through each fuse matches closely so as to balance the current in the two fuses as much as possible. An alternative fuse that does not need paralleling, but is harder to mount, is the Bussman JJS-150.[5]

During charge, the midpack contactor, HVRL1, is activated, but the Toyota contactors, SMRL1–3, are not, thereby keeping the battery pack isolated from the hybrid system. The Toyota contactors are locked out whenever the vehicle is being charged (for details, see the control board description), keeping the battery pack isolated from the hybrid system.[6]

Contactors

When the Prius is running (READY light), all three contactors—HVRL1, SMRL2, and SMRL3—are activated, connecting the full pack voltage to the hybrid system. To reduce inrush currents, a precharger circuit is used. The stock hybrid system first connects the battery to the hybrid system through SMRL1, a resistor, and SMRL3. Then SMRL2 is activated and SMRL1 is deactivated.

Toyota-Related Controls[6]

I have included, on and off the control board, circuitry to control battery cooling fans. These can be controlled through a computer using any criteria you choose. However, if you can redirect the OEM battery's fan to your battery pack, this has the advantage of proportional control based on ambient and battery temperatures. You will need to replace Toyota's battery thermisters with units mounted at strategic locations that are in thermal contact with your battery pack. You may need to fudge their values or range to relate your battery's acceptable temperature range to that of the OEM battery, which is programmed into the battery ECU. The NiMH voltage/temperature coefficient is probably also programmed in, and should be taken into consideration. In the right rear fender well is an exhaust port that can also be reused. If you use your own fan and control, it is recommended that air still be drawn from the passenger compartment and exhausted to the outside.

Note that operations at very low and very high temperatures have not yet been verified or planned for. In fact, my PRIUS+ needed to be switched back to its OEM battery pack before it would operate at all in freezing temperatures. There are potential issues of non-NiMH battery packs having different temperature coefficients from those that the battery ECU is programmed for.

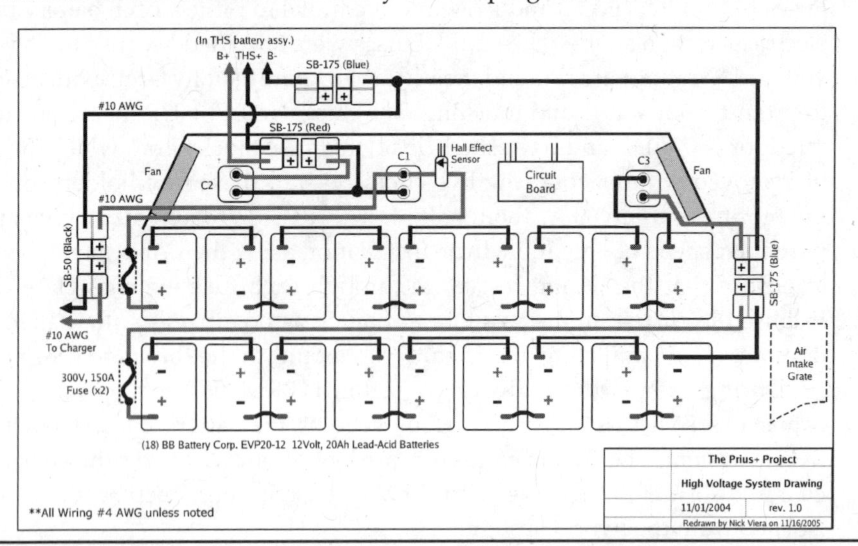

FIGURE 10-3 HV system layout and wiring diagram of CalCars' first PRIUS+. Originally hand drawn by Ron Gremban, then converted to CAD and PDF by Nick Viera.

The battery ECU uses *voltage taps* from the OEM battery pack to verify battery integrity and charge balancing. The 168-cell battery is thereby electrically divided into fourteen 12-cell subpacks (each subpack consists of two 6-cell prismatic modules).

Battery Tap Emulator Board

Even if your battery pack is similarly divisible, you will need to use a resistive divider like the battery tap emulator board (as seen in Figure 10-4) to simulate these battery taps for the battery ECU. Once the taps are emulated, the battery ECU can be fed any voltage as its perceived battery voltage. Battery ECU state-of-charge spoofing is done by raising this perceived voltage above the actual battery voltage. Once you have emulated the battery taps, you will need to provide this as necessary for your battery pack's charge balancing and actual SOC management. As this can be done in many ways for the many battery pack possibilities, it is outside the scope of this document.

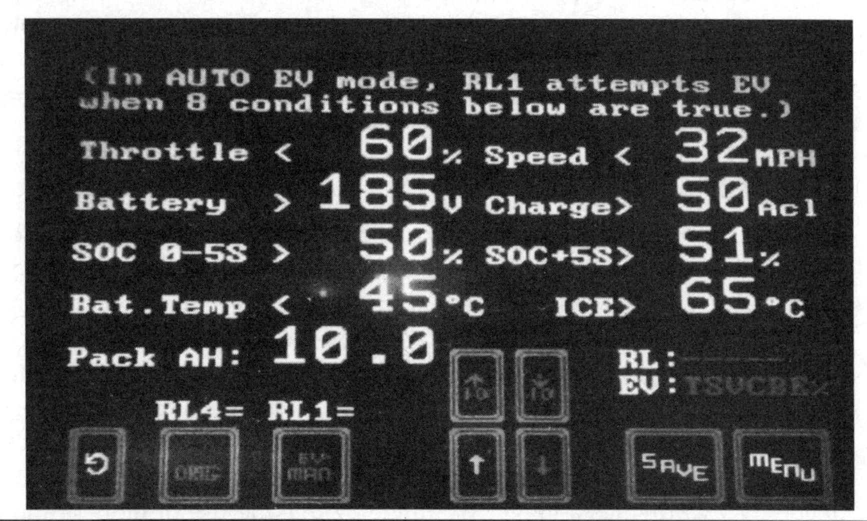

FIGURE 10-4 Battery tap emulator 2g.

TABLE 10-2 Notes for Figure 10-4

Setting	Notes
Throttle < 60% Speed < 32MPH Battery > 185v Charge > 50Acl SOC 0–5s > 50% SOC+5s > 51% Bat.Temp < 105°F ICE > 105°F Pack AH: 10.0 (adjust per PHEV pack)	These are the settings found on the Setup –> PHEV page. These eight options dictate the manner in which the Auto EV-Mode functions work. RL1 is used to trigger the EV-Mode button, which can be triggered manually or by the Auto EV-Mode function. The SOC 0–5s value is used during the first 5 seconds when the vehicle is powered on; the SOC+5s value is used after startup during normal driving. There is an EV Man/EV Auto button on the menu bar of each of the normal screens which disables or enables the Auto EV-Mode function.

Control Board

The control board[8] (low-power control schematic; see Figure 10-5) provides the overall logic for a PRIUS+. It has a drive interlock and a charge interlock relay. The charge interlock relay can be jumped to be energized whenever the charger is connected to the battery pack or, if the charger is built in and always connected, when the charger is powered. The charge interlock relay prevents driving while charging by disengaging HVRL1 and thereby disconnecting the two halves of the battery pack. The drive interlock relay energizes HVRL1 only when the vehicle is in READY mode.

TTL inputs to J3 control the activation of the EV-only mode, fans, and/or BMS spoofing via opto-isolators. An isolating power supply supplies 30 V, an adjustable portion of which can be added to the battery voltage when necessary to provide battery management system SOC spoofing. These TTL levels can be provided from a computer's parallel port, independent of the computer's power supply. Battery voltage is applied at HVJ6. Additional voltage, adjustable via R7, is added when RL2 is energized, and the resulting voltage is applied at J8, where the battery tap emulator board plugs in.

Battery ECU SOC Spoofing

Note: This is a work in progress. Spoofing requires monitoring the battery ECU's perceived SOC, available on the CAN bus. The idea is to keep the battery ECU's perceived SOC high (above 70 percent) until the battery pack's real SOC reaches a value that I will call normal hybrid SOC—the point where battery depletion (average discharge or PHEV mode) ceases and normal hybrid operation takes over. Normal hybrid SOC is determined by first deciding what you want to use as your battery pack's deepest depth of discharge (DOD) during normal operation (the battery should only rarely be discharged further than this). This is a trade-off, as a deeper DOD provides a longer EV range by making use of more of the battery's capacity, but a shallower DOD gives the battery a longer cycle life. An 80 percent DOD (discharge to 20 percent SOC) is common. For the Prius, the normal hybrid SOC needs to be about 1 ampere-hour (Ah) above maximum normal DOD; this is because the Prius's hybrid system's normal operations occur within about 1 Ah either side of 60 percent (perceived) SOC. As an example, imagine my PRIUS+'s EVP20 battery pack has an effective capacity of 15 Ah. An 80 percent DOD would be at 20 percent (real) SOC, 3 Ah up from full discharge or 12 Ah below full charge. Normal hybrid SOC would then be 4 Ah up from full discharge, 11 Ah below full charge, or 4/15 = 27 percent (real) SOC. Start with a spoofing voltage of, say, 12 V. The spoofing algorithm still requires some fine tuning, as it was developed with lead-acid batteries and no automation, but it generally goes as follows:

- When the battery is full and can accept little regenerative braking current, energize the added spoofing voltage whenever the perceived SOC is 81

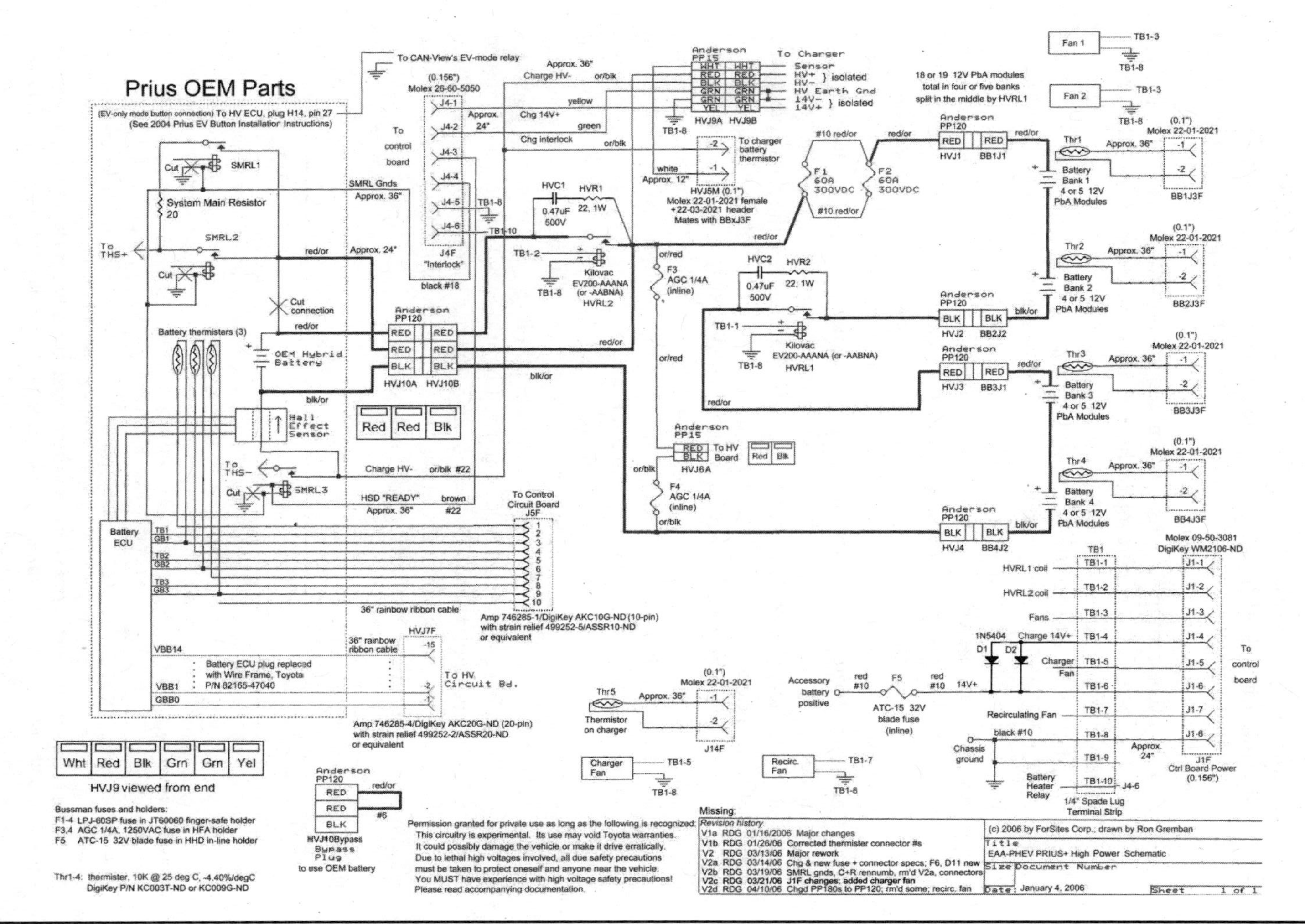

Figure 10-5 Control schematic 2f. Source: EAA-PHEV.

percent or lower, then remove the spoofing voltage whenever it reaches, say, 82 percent. As the perceived SOC may keep moving in the previous direction for a short time after a change (overshoot), you may find that to maintain ≥80 percent SOC, you have to switch the spoofing voltage on at even higher than 81 percent.

- Regenerative braking current is limited to around 20 A whenever the perceived SOC is 80 percent or higher. However, sometimes the engine races when the SOC is set this high, or it becomes impossible to enter EV-only mode. If either of these things happens, reduce both SOC endpoints until it doesn't happen anymore. If the engine starts racing, you can clear it by putting the car in neutral, then back in drive.
- As long as the perceived SOC is 70 percent or higher, the hybrid system tries to use up "excess" battery charge, so once the battery can accept regenerative braking, add spoofing voltage whenever the perceived SOC is as low as 74 percent; remove it whenever the perceived SOC reaches 78 percent. This allows for some overshoot, as explained earlier. Change the endpoints as necessary to keep the perceived SOC within the 70–79 percent range.
- As the battery pack gets more and more discharged, at some point, perceived SOC will decrease below 70 percent even with the spoofing voltage turned on. At the point where the battery pack's real SOC (usually determined by ampere-hour integration) reaches its normal hybrid SOC (explained at the beginning of this section), perceived SOC with spoofing voltage turned on should reach 60 percent (the SOC that the hybrid system tries to maintain). If the battery pack has too much charge left (real SOC > normal hybrid SOC) when the perceived SOC reaches 60 percent (with spoofing voltage turned on), adjust R7 to raise the spoofing voltage; if the battery pack has too little charge left (real SOC < normal hybrid SOC) when the perceived SOC reaches 60 percent (with spoofing voltage turned on), adjust R7 to lower the spoofing voltage. Getting the right spoofing voltage may require many test runs, each starting with full batteries.

Reference Materials

Table 10-3 Disclaimer

Permission granted for private use as long as the following is recognized: This circuitry is experimental. Its use may void Toyota warranties. It could possibly damage the vehicle or make it drive erratically. Due to lethal high voltages involved, all due safety precautions must be taken to protect oneself and anyone who comes near the vehicle. You MUST have experience with high voltage safety precautions as referred to in the Main Disclaimer of this book and the EAA-PHEV!
Note: This modification can add 200 to 300 pounds of extra weight in the rear of the vehicle, depending on the specific battery pack used and whether or not the OEM battery is removed. This will use up some of the Prius's available rear axle useful load. In practice, my PRIUS+'s handling has been fine, although it rides a little low in the rear. I recommend avoiding carrying rear passengers and significant load in the cargo area simultaneously.

Source: EAA-PHEV; Ron Gremban, 1/18/2006. © 2006 by ForSites Corp. in coordination with EAA-PHEV.

Installing the CAN-View[9]

Installing the CAN-View and Prius EV-mode button both involve disassembling the same areas of the Prius, so they will be covered here at the same time. You should also be familiar with these reference materials.

Disassembly of the Dashboard

We begin by disassembling the dashboard.

1. Remove the bottom cover of the steering column.
   a. Release the steering wheel adjustment handle and remove the silver screw.
   b. Turn the steering wheel 90 degrees to right and left to remove the black screws on each side.
   c. Remove the lower cover by carefully pulling down on the lower half.
2. Disconnect the headlamp flasher plug on the left side of the steering column.
   a. Remove the plug's cover by unlocking the two side tabs/clips.
   b. Remove three unused connector pins *using a jeweler's screwdriver to disengage and slide them out.*
   c. Optionally, to use the headlamp flasher circuit as an EV-mode button:
      - Snap the cover back onto the flasher plug and reinsert it into the switch.
      - Reassemble the steering column, being sure everything is aligned properly.
3. Remove the lower glove box by squeezing the inside sides together to lower the box below the catches, then unclip the small piston from the right side. Lower the box until the lower joints detach from the dash and remove the pneumatic cylinder, noting its orientation.
4. Remove the passenger-side silver air vent cover by pulling it out from the bottom first. Next, remove the small interior colored piece just below the vent piece by pulling it straight out.
   a. If you are installing CAN-View Without Navigation, refer to www.eaa-phev.org/wiki/PiPrius_conversion_process#CAN-View_Ref3.
   b. Remove the driver's-side air vent cover by pressing down and pulling it out on top.
   c. Remove the lower center console hump with the 12-V lighter power socket, remove barb from passenger side, and pull out.
   d. Remove the air vent at the right side of the MFD screen; open the upper glove box to pull out the vent.
   e. Remove the lower driver's-side interior colored dash panel; there is one black screw above the hood release and one exposed behind the driver's-side vent.

f. Remove the black lower dash key fob panel, and leave it hanging with the wires connected.
g. Remove the upper driver's-side dash panel with power button; leave wires attached.
h. Remove the air vent at the left side of the MFD screen; with the shift lever remaining in place, detach and slide the park button forward through the silver panel to expose and detach the cable, then remove the panel and reattach the park button to its cable to prevent errors during later tests.
i. Remove 10-mm bolts, one on each side of the MFD screen, pull the screen out sharply, and rotate it toward the driver's side.
j. Tap the gray OEM wire for 12-V power (*top row, second from the left*).
k. Attach the CAN-View video cable to the MFD and power spade to the tapped gray wire; run the cable out directly behind the screen.
l. Connect the CAN-View OBDII cable to the OBDII port and route the cable behind the center console toward the glove box along with the headlamp flasher EV-mode button wire if installed.
m. Test CAN-View, then reassemble the center and driver's side of the dash.
n. CAN-View will be mounted under the passenger seat, above the JBL amplifier if one is present.

5. If installing CAN-View With Navigation:
   a. Route the OBDII cable around the foot well and down the driver's-side door sill to underneath the driver's seat. Route the headlamp flasher EV-mode button wire, if installed, behind the center console and glove box.
   b. Tap the gray OEM wire from the navigation unit for 12-V power.
   c. Connect the CAN-View video cables to navigation unit.
   d. Test and attach above and to the rear of the navigation unit.
6. Solder the new pin to the black wire and optionally to the black wire that connects to the headlamp flasher.
7. Install the pin with black wire(s) to the EV-mode button location in the HV ECU *H14#27*.
   a. The HV ECU is the one with gray plugs closest to the exterior of the car.
   b. H14 is the lowest of the four connectors.
   c. Pin 27 is located on the most interior (broken into three segments) row, the second from the bottom left corner, in the only open location between two red wires.
   d. Using a jeweler's screwdriver, raise the white terminal retainer, fully insert the new pin, recompress the retainer, and plug the terminal back into the HV ECU.
8. Route the black EV-mode wire, and the OBDII and video if the installation is without navigation, along the passenger-side door trim, exiting under the carpet before the pillar to the hole in the carpet below the passenger seat and to the relay cable.

9. Route the relay cable from CAN-View under the rear passenger-side door trim toward the rear of the car.
10. Reassemble the passenger side of the dash and door trim.

Finally, connect all the cables to CAN-View.

Simple CAN Bus Scanner

Figures 10-6 to 10-8 show a few photos of Jim Fell's simple CAN bus controller for the Prius. It just reads the three messages of relevance from the OEM battery pack and can process any control algorithm necessary, then send the signals to Rich Rudman's PFC.

CAN-View Version 4

Another set of instructions may be needed for the CAN-View Version 4, as it does not integrate with the OEM MFD, but gets power directly from the OBDII port. Prius CAN-View V4 mounting options at PriusChat.com covers various ways of mounting the second touch screen.

CAN-View Configuration

The CAN-View should be preconfigured, but if you update the firmware or change the settings and forget the defaults, here is what we use for the PiPrius conversion settings. Please refer to the text settings as opposed to the screen shot images, as the photos may not be as up to date. The bold text represents the changeable elements that the user can customize. Please feel free to use slightly different values, depending on your unique driving conditions or the characteristics of your particular vehicle. Some of the general settings we change from the defaults are Sound: **no**; Autosave: **no**; Info: **start**; Black background: **yes**: **@night**. Be aware that you must use the Save button in order for changes to take effect after the car is powered off. It is recommended that you save on the page that you would like to have be the default page you see when the car is powered on.

FIGURE 10-6 Prius PHEV user interfaces—simple CAN bus scanner created by Jim Fell.

FIGURE 10-7 Felix Kramer's instrumentation on the Prius PHEV, installed by CalCars.

FIGURE 10-8 Ford Escape PHEV instrumentation.

PHEV Page 1

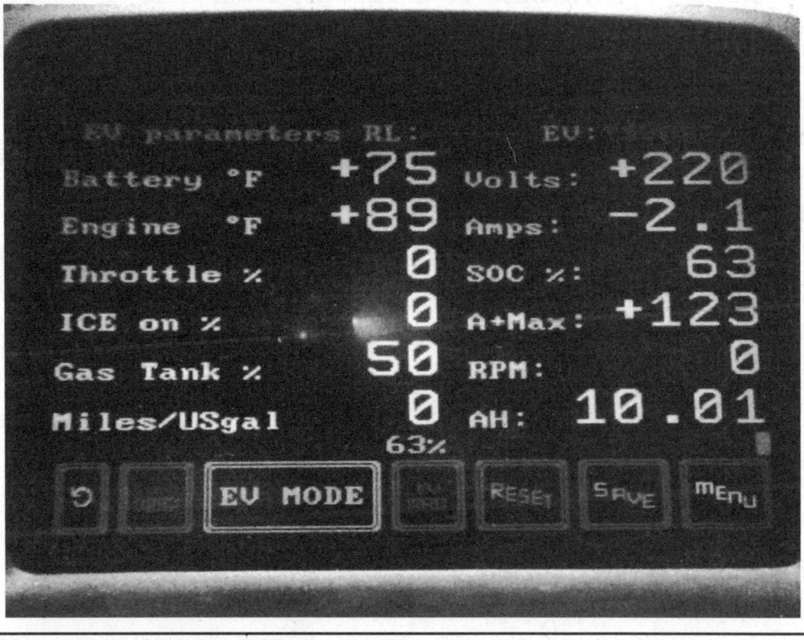

FIGURE 10-9 PiPrius_CAN-View_PHEV1.

FIGURE 10-10 Ryan installing CAN bus software changes.

TABLE 10-4 PHEV Page 1 Setup Information

Battery Temp	+75°F	The left column of values can be changed while the right column is static. These values are simply suggestions which seem to work best for the PiPrius development team but you are certainly free to change them as you see fit. Note: We suggest that you clear your trip page values, press the AH counter to reset it's value, and then return to this page and save whenever you make configuration changes. This will ensure that each trip will start with 0 distance and time and that the AH counter begins from the same state on each trip.
ICE Temp	+89°F	
Throttle %	0	
ICE On %	0	
Gas Tank %	50	
Miles/USGal	0	
Volts	+220	
Amps	–2.1	
SOC %	63	
A +Max	+123	
RPM	0	
AH	10.01	

PHEV Setup Page

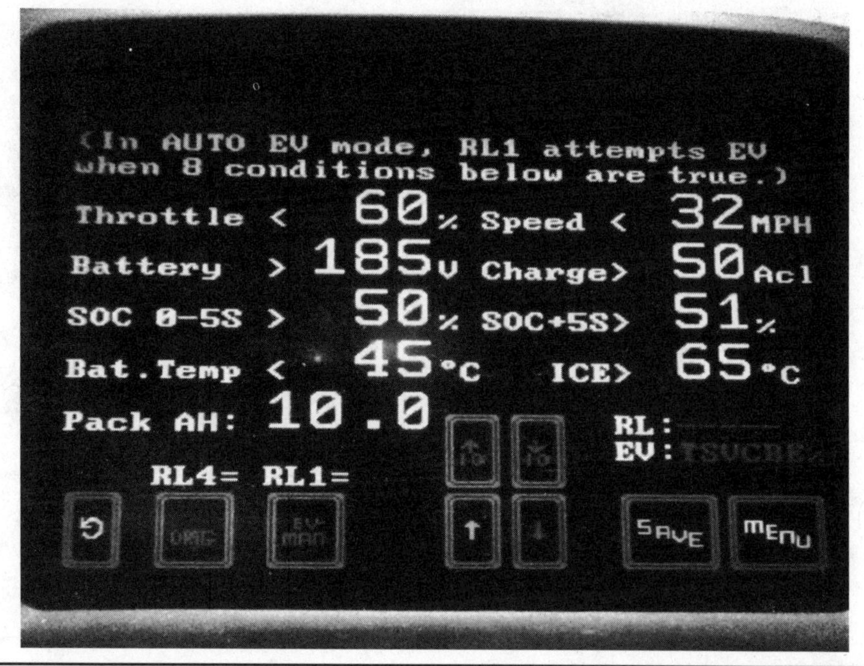

FIGURE 10-11 PHEV setup page.

TABLE 10-5 PHEV Setup Page

Throttle < 60% Speed < 32MPH Battery > 185v Charge > 50Acl SOC 0–5s > 50% SOC+5s > 51% Bat.Temp < 105°F ICE > 105°F Pack AH: 10.0 (adjust per PHEV pack)	These are the settings found on the Setup -> PHEV page. These eight options dictate the manner in which the Auto EV-Mode functions work. RL1 is used to trigger the EV-Mode button, which can be triggered manually or by the Auto EV-Mode function. The SOC 0–5s value is used during the first 5 seconds when the vehicle is powered on; the SOC+5s value is used after startup during normal driving. There is an EV Man/EV Auto button on the menu bar of each of the normal screens which disables or enables the Auto EV-Mode function.

RL2 Setup Page

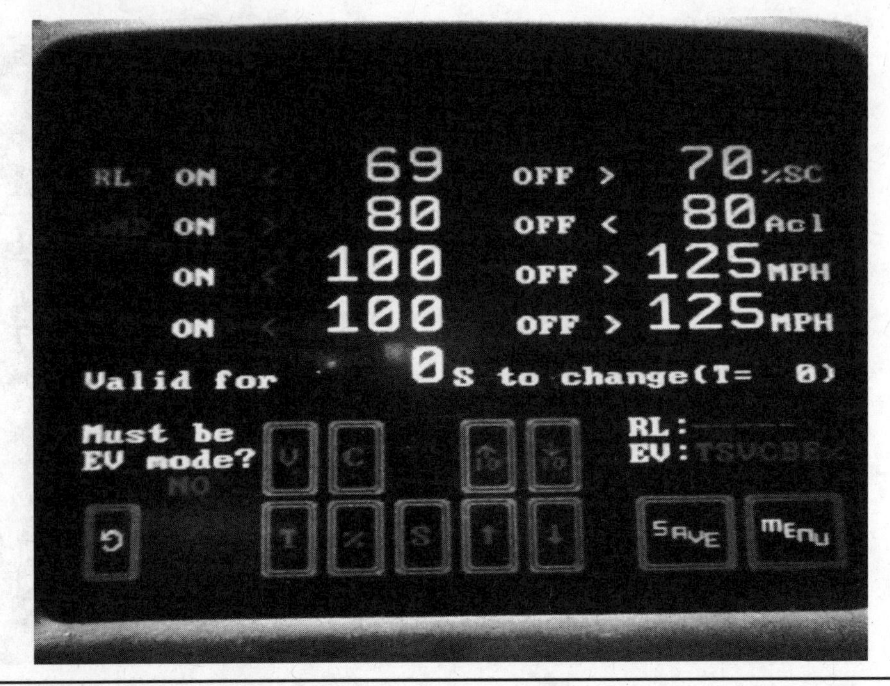

Figure 10-12 RL2 setup page.

Table 10-6 RL2 Setup Information

All values are AND rather than Red. ON < 63 OFF > 65 %SC ON > 80 OFF < 80 Acl ON > 100 OFF < 125 MPH (default) ON < 100 OFF > 125 MPH (default) Valid for 0s to change Must be in EV mode? NO (default)	This relay is used to activate the PFC Charger's High NiMH Voltage setting, which will cause the SOC to rise due to State Of Charge Drift. The Acl (Amps Charge Limit) is monitored to keep from overcharging the NiMH and attempts to leave some headroom for regen power. Unlike RL3, speed is not monitored so this relay's lower SOC target is used during low speed situations when there will typically be more regen and EV-Mode can be taken advantage of.

RL3 Setup Page

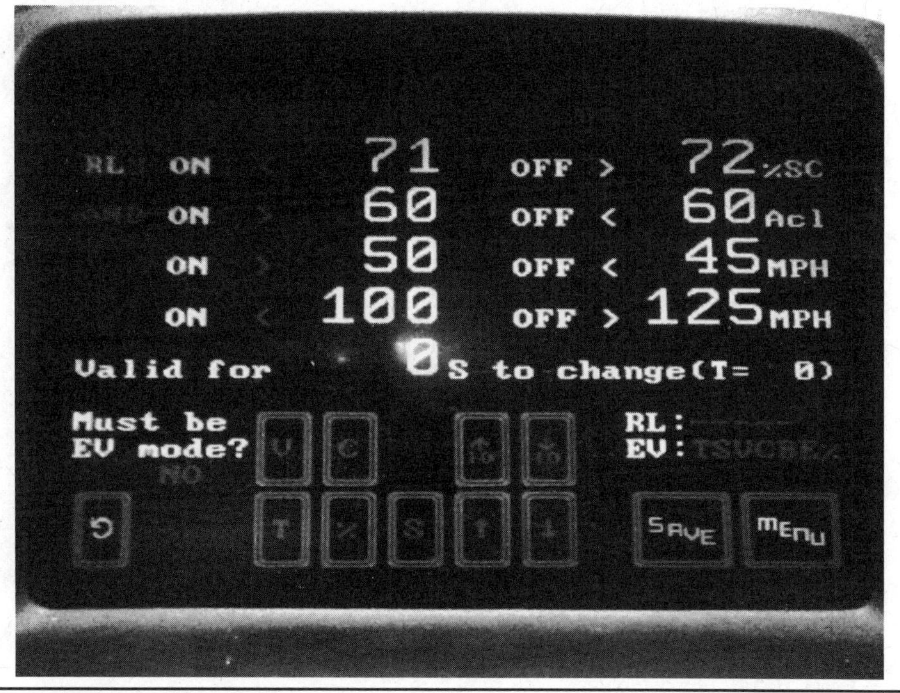

FIGURE 10-13 RL3 setup page.

TABLE 10-7 RL3 Setup Information

All values are AND rather than Red. ON < 72 OFF > 73 %SC ON > 90 OFF < 90 Acl ON > 50 OFF < 45 MPH ON < 100 OFF > 125 MPH (default) Valid for 0s to change Must be in EV mode? NO (default)	This relay is used to activate the PFC Charger's High NiMH Voltage setting, which will cause the SOC to rise due to State Of Charge Drift. The Acl (Amps Charge Limit) is monitored to keep from overcharging the NiMH and attempts to leave some headroom for regen power. The speed is monitored such that the higher SOC target will be used while on the freeway to enable better Mixed-Mode operation.

RL4 PHEV/OEM Mode

RL4 is used to enable the PHEV mode or revert back to OEM operation. There are no configuration settings for this relay—only the **orig/phev** button from the menu bar that enables or disables the PHEV system. If the OBDII cable is ever disconnected, the CAN-View will automatically disable the PHEV system.

RL5 Setup Page

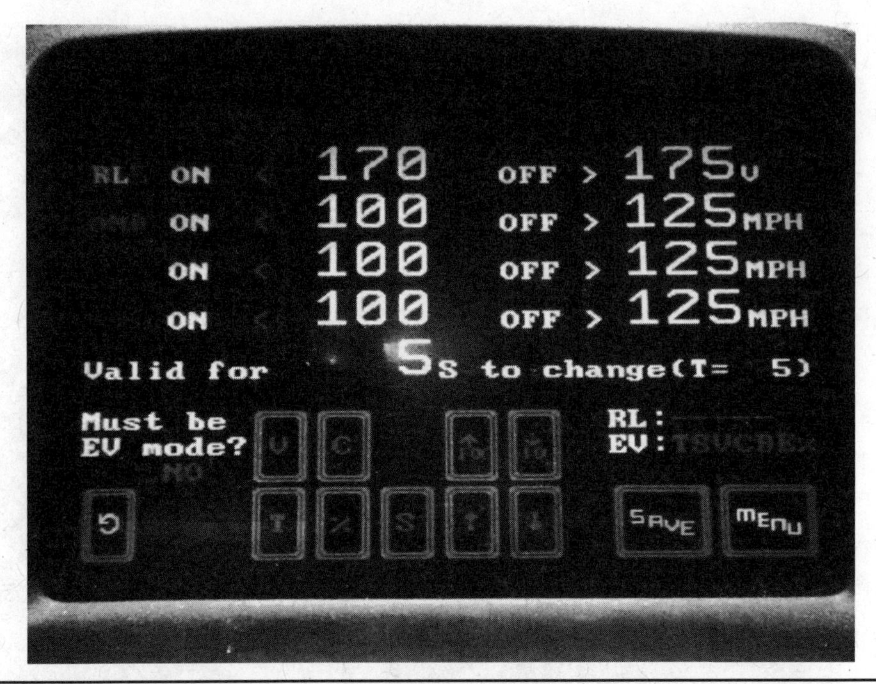

Figure 10-14 RL5 setup page.

Table 10-8 RL5 Setup Information

All values are ANDed rather than ORed. ON > 98 OFF < 98 B°F ON < 100 OFF > 125 MPH (default) ON < 100 OFF > 125 MPH (default) ON < 100 OFF > 125 MPH (default) Valid for 0s to change Must be in EV mode? NO (default)	This relay is used to bypass the OEM control over the NiMH battery cooling fan. The OEM battery does not function to it's full capabilities until it warms up into the 90°Fs, however if it is allowed to reach ~105°F then EV-Mode is denied. The OEM fan is more than capable of rapidly cooling the OEM battery however the Toyota system rarely allows the fan to run at full speed, in fact we have seen it reach as much as 130°F without much demand for additional cooling from the fan.

RL6 Setup Page

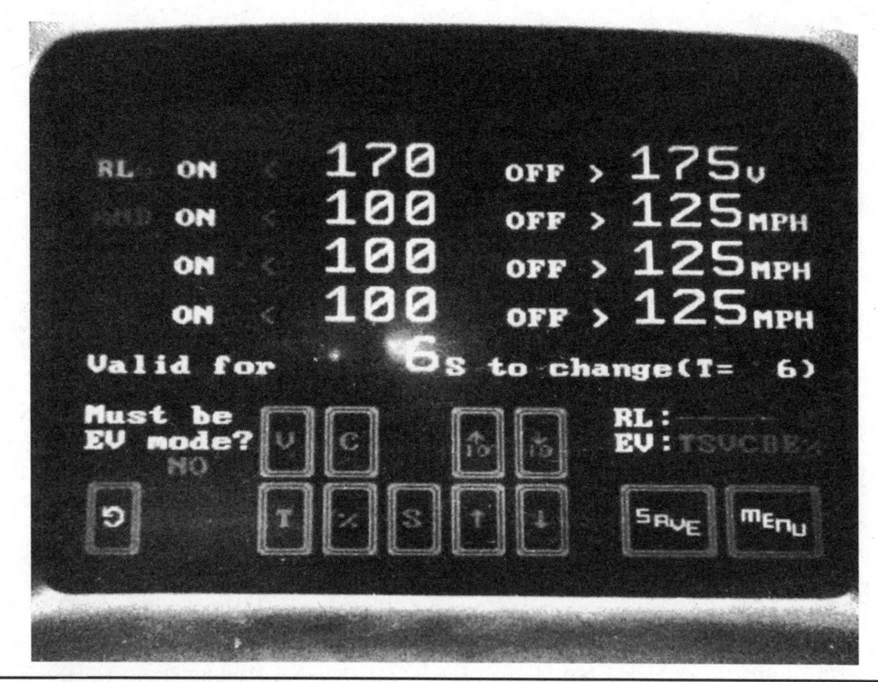

FIGURE 10-15 RL6 setup page.

TABLE 10-9 RL6 Setup Information

All values are AND rather than Red. ON < 170 OFF > 175 V (default) ON < 100 OFF > 125 MPH (default) ON < 100 OFF > 125 MPH (default) ON < 100 OFF > 125 MPH (default) Valid for 6s to change (default) Must be in EV mode? NO (default)	This relay is not currently in use, so it has been left with the default values. The first option is always false, while the remaining three are always true. These defaults allow for ease of configuration, because upon changing the first value to some useful settings, the remaining always true values can be ignored.

OEM HV Battery Modifications

Removal of the Batteries

From the hatch opening:

- Remove the carpet in the hatch compartment floor to reveal the battery.
- Turn the orange safety plug from LOCK to UNLOCK and pull it out.
- Remove the black plastic air coupling on the rear left.
- Remove the bolts on either side of the battery (three bolts on each side).
- Lift the bottom of the rear right passenger seat and move it forward.
- Lift the strip of carpet to reveal the metal cover over the high-voltage cables.
- Remove the two (not three) nuts holding the black metal cover.
- Flip the seat forward to see the other end of the black metal cover.
- Remove the two bolts holding the other end of the black metal cover to the battery.

Disconnect the Battery

- From the rear right seat, remove the orange HV connector on the right (flip the lever).
- From the rear left seat, remove the big black signal connector on the left (unbolts with a 10-mm socket wrench).
- From the rear left seat, remove the small connector next to the signal connector (snaps).

Remove the Battery

- Remove the six bolts, three on each side of the battery, bolting it to the floor (.5" socket).
- Hook an engine hoist to the two round holes in the black metal on either side of the battery.
- Hoist the battery out of the car.

Open the Battery

- You need a #35 security Torx driver and a #35 Torx driver.
- Remove all the screws in the two top covers:
  - The cover over the fans
  - The cover over the batteries and electronics

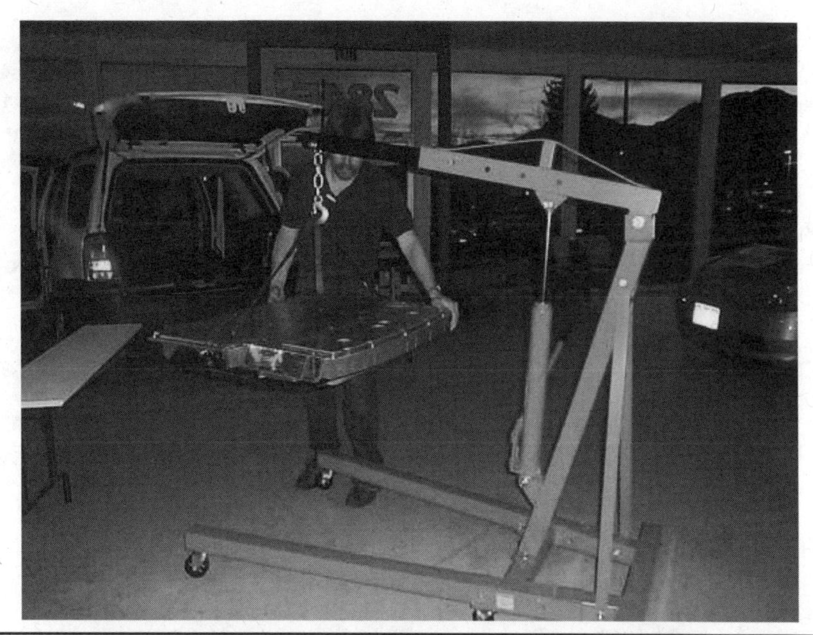

Figure 10-16 Battery pack lifted out of the Ford Escape Hybrid electric vehicle.

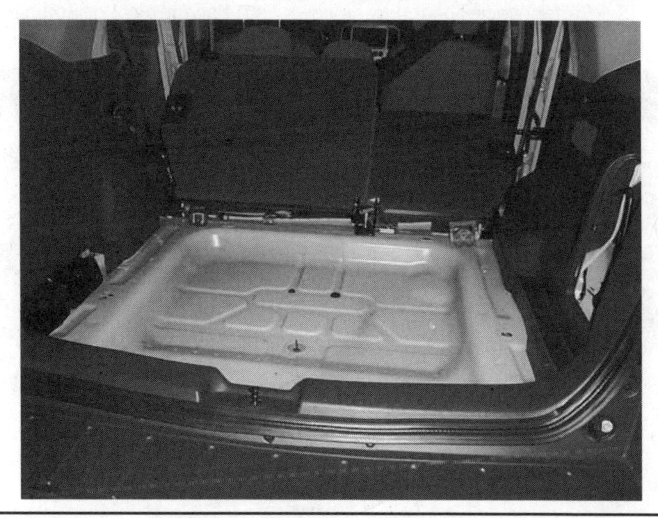

Figure 10-17 Ford Escape, trunk without the battery.

FIGURE 10-18 Battery pack opened (rear view), Ford Escape Hybrid. Source: EAA-PHEV.

FIGURE 10-19 Battery pack opened (front view), Ford Escape Hybrid. Source: EAA-PHEV.

Component Locations

The battery includes (when in the vehicle)

- Air blowers in the rear compartment.
- NiMH cells in the center.
- Two layers of cells.
- Each layer has a left and a right group.
- The groups in the top layer have 13 columns of 5 cells in series.
- The groups in the bottom layer have 12 columns of 5 cells in series.
- Total: $2 \times 13 \times 5 + 2 \times 12 \times 5 = 250$ cells.
- Nominal pack voltage: $1.2\ \text{V} \times 250 = 300\ \text{V}$.
- Controller is on the right side.
- Contactors and HV connector are in the right front corner.
- HV safety plug is in the right rear corner.

The current sensor is inside the converter on the left side.

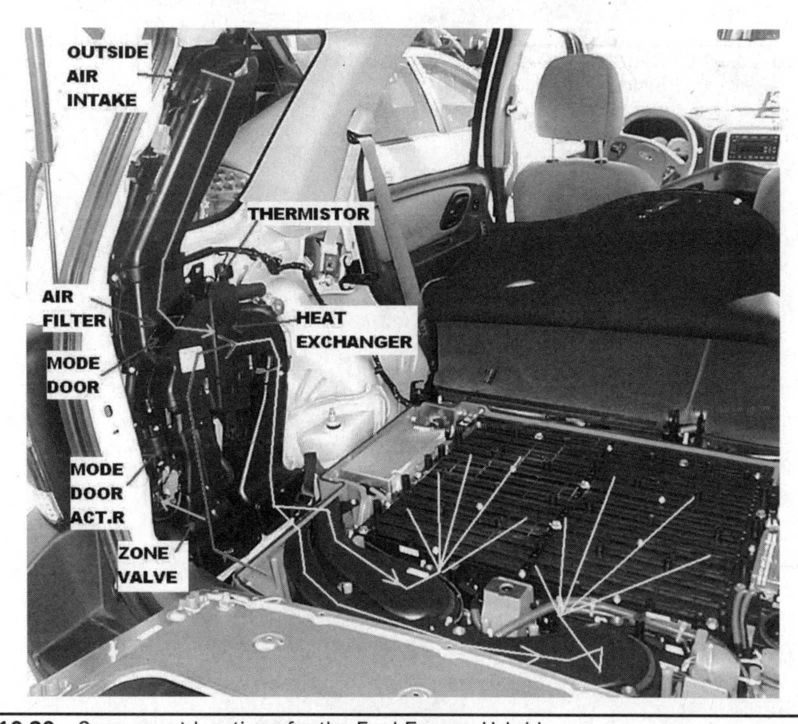

FIGURE 10-20 Component locations for the Ford Escape Hybrid.

Names and Functions

Notes to Table 10-10:

- Green, OK: function is understood and confirmed.
- Red, ??: function is not understood or not yet confirmed.
- Gray, n.a.: PHEV conversion can work without this function.

Battery Cooling System

FIGURE 10-21 Battery HVAC air flow, exhaust, forced air flow, outside air intake, and exhaust inside vehicle.

FIGURE 10-22 Battery HVAC shown on a Ford Escape Hybrid.

TABLE 10-10 Names and Functions

Group	Ckt	Pin(s)	Ext. color	Int. color	Name	Dir	Function	Notes	
12V pwr	57	35,36, 37	LtgrnBlk	Blk	Ground	IN	Power ground		OK
	570	30,31	BlkWht	BlkWht	Ground	IN	Signal ground		OK
	3800	4,5,6	LtgrnBlk	Red	+12 V	IN	Power +12V	Always on	OK
	16	10,11	RedLtgrn	RedBlu	+12 V	IN	Low power +12V	Always on	OK
	3206	19	LtgrnYel	TanRed	Voltage supplied in Start and Run	IN	Receives 12V when the ignition switch is in either the On or Start positions (PHEVen if engine is not running)	From the ignition switch. Overload protected	OK
	3997	14	Dkgrn	Tan	Power sustain relay out	IN	Receives 12V when the ignition switch is in either the On or Start positions (PHEVen if engine is not running) and for 2 seconds after the ignition is turned off	Fed by the Powertrain Control Module's Power Relay, located in the Battery Junction Box. The Powertrain Control Module is located under the hood, in the rear-center	OK
Air intake	3703	21	BrnWht	BlkBrn	Battery compartment thermistor signal	IN	Senses air intake temperature	All are located inside the column at the rear-left corner of car, inside air intake ducts	NA
	3704	25	DkgrnWht	WhtBlk	Battery compartment thermistor return	IN	Senses air intake temperature	All are located inside the column at the rear-left corner of car, inside air intake ducts	NA
	698	34	Red	RedBlu	Mode door actuator motor +	OUT	Moves a flap controlling air flow	All are located inside the column at the rear-left corner of car, inside air intake ducts	NA
	699	26	Org	BlkYel	Mode door actuator motor –	OUT	Moves a flap controlling air flow	All are located inside the column at the rear-left corner of car, inside air intake ducts	NA
	1129	17	BrnWht	RedGRn	Mode door actuator potentiometer +	OUT	Senses position of flap	All are located inside the column at the rear-left corner of car, inside air intake ducts	NA
	1130	20	PnkLtgrn	BluBlk	Mode door actuator potentiometer wiper	IN	Senses position of flap	All are located inside the column at the rear-left corner of car, inside air intake ducts	NA
	1128	24	GryLtBlu	BlkWht	Mode door actuator potentiometer –	OUT	Senses position of flap	All are located inside the column at the rear-left corner of car, inside air intake ducts	NA
	698	34	Red	RedBlu	Zone Valve	OUT	Solenoid selecting air source	All are located inside the column at the rear-left corner of car, inside air intake ducts	NA

Table 10-10 Names and Functions *(continued)*

Group	Ckt	Pin(s)	Ext. color	Int. color	Name	Dir	Function	Notes	
CAN bus	1908	29	Wht	YelRed	High speed CAN bus +	I/O	Communicates with vehicle	See CAN section below for messages	OK
	1909	28	Blk	YelWht	High speed CAN bus –	I/O			OK
Jump start switch	176	16	PnkLtgrn	BrnWht	Jump start switch feed	IN	When grounded, lets 12V battery jump charge-up the traction battery a bit, through DC-DC converter in battery pack, enough to start the car	The switch is located to the left of the driver's left ankle, behind a black plastic panel	OK
	179	12	OrgRed	GrnBlk	Jump start switch illumination +	OUT	When at 12V, it lights-up the switch		OK
Emergency control	3003	8	VioWht	Tan	Battery power off signal	OUT	0–12V square wave, 50% duty cycle. If all OK, 2 Hz. If problem, 6 Hz. From the Traction Battery to the Power Train Control Module	The Power Train Control Module is located under the hood, in the rear-center	OK
	877	7,23	Wht	RedBlk	Fuel pump feed / Inertia Sw input	IN	Normally receives 12V when the ignition switch is in either the On or Start positions (PHEVen if engine is not running) and for 2 seconds after the ignition is turned off; no voltage when the ignition is off, or in case a crash opens an inertia switch	The High Voltage Cutoff switch is located in the right-rear column of the car	OK
	212	27	Dkblu	BlkBlu	Immediate shutdown 1	OUT	The Traction Battery tells the Transaxle Control Module that all is OK by sending 12V (same duration as the Sustain line). If both lines are open, the Transaxle Control Module starts a fault	The Transaxle Control Module is under the hood, in the center, to the left of the box labeled "HYBRID"	OK
	213	13	DkbluYel	BlkRed	Immediate shutdown 2	OUT			OK
Unused	n.a.	18	n.a.	TanRed	???	???	???	Connected to controller, not used in vehicle	?
		32		YelBlk					?

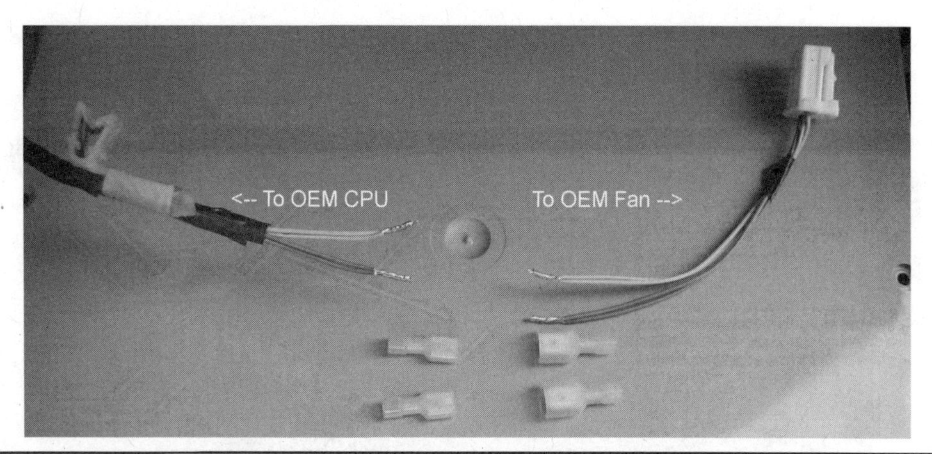

FIGURE 10-23 Wire connections for fan control on Toyota Prius PHEV conversion.

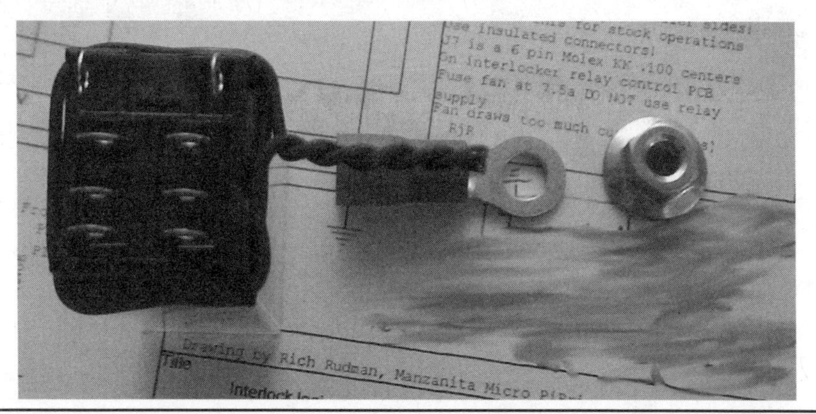

FIGURE 10-24 Connectors for fan control on Toyota Prius PHEV conversion.

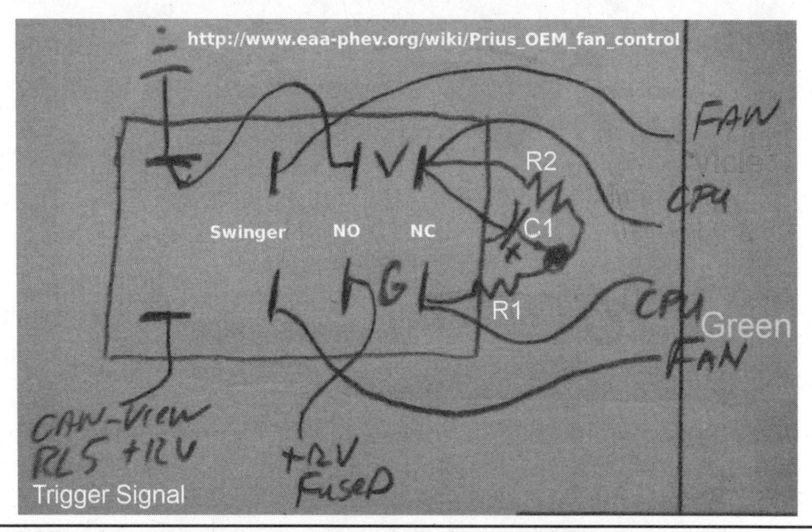

FIGURE 10-25 Wiring diagram for fan control on a Toyota Prius PHEV.

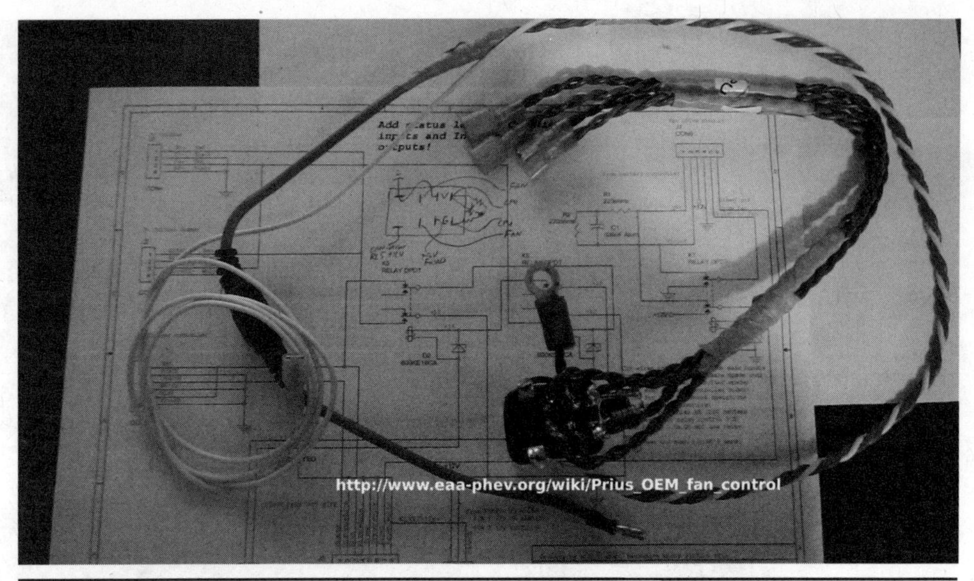

FIGURE 10-26 Completed wiring for fan control on a Prius PHEV conversion.

Battery HVAC

The cooling system controls the temperature of the NiMH cells in the traction battery.

Its components are located

- In the rear left column
- In the traction battery itself

When the system is used in a closed loop, air flows

- From the empty spaces in the battery pack
- Out of the rearmost grille in the rear right corner of the battery pack
- Into a duct in the rear right column of the vehicle up the rear duct in that column
- Through the mode door that controls the air flow (unconfirmed)
- Through the zone valve that selects the air source (unconfirmed)
- Forward through a heat exchanger
- Down the front duct
- Out of the rear right column
- Into the frontmost grille in the rear right corner of the battery pack
- Into the battery pack
- Into two ducts, one for each blower
- Into two blowers, one for each duct

- Into each set of cells (left set for the left blower, right set for the right blower)
- Through the cells and into the empty spaces in the battery pack
- Completing the cycle

The heat exchanger is chilled by the vehicle's air conditioning system. This is done through two metal pipes, which run from the bottom left corner of the vehicle, then forward, behind (to the left of) the black plastic ducts, and up to the heat exchanger. Condensation collected in the heat exchanger flows into two rubber tubes below it, through a Y into a single rubber tube, and through the floor to let the condensation drip on the ground.

When using outside air:

- Air is taken from a vent in the rear right window
- Down a duct
- Through an air filter
- Through the mode door
- Into the heat exchanger
- Then following the same path as described previously

Now that extra air has been taken into the system, air has to be let out of it:

- Air from the pack flows into the rear duct.
- The zone valve opens, letting air from the rear duct out into the open space in the rear right column.
- From there, air flows into the rear storage area.

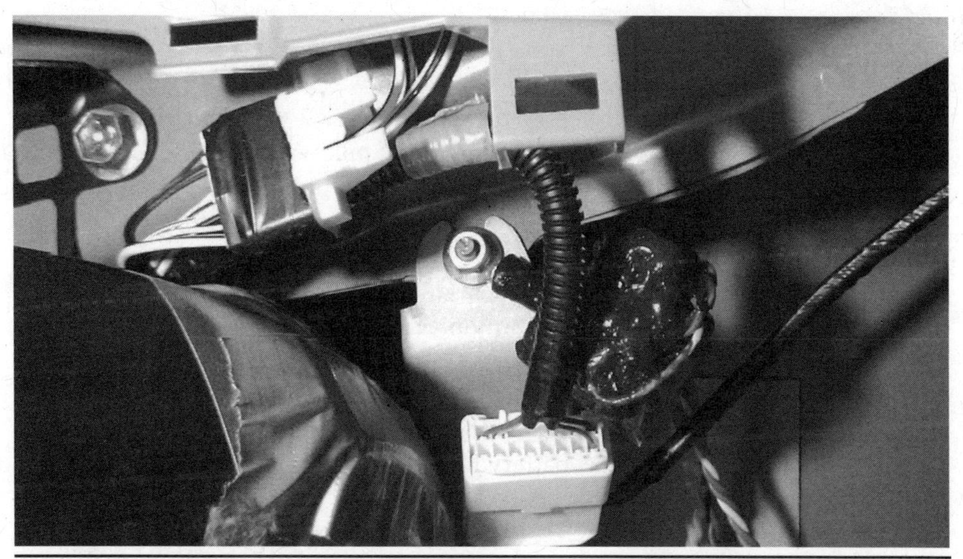

Figure 10-27 Connectors for the Prius fan control.

To monitor the temperature, thermistors are placed as follows:

- One in the rear right ducts
- Two by the blowers, one for each blower
- Two in the left block of cells and two in the right block (unconfirmed)

To control the air flow and temperature:

- The blower's speed is variable.
- The mode door's position is continuously variable from closed to fully open.
- The zone valve selects the air source.

The vehicle varies the blowers' speed based on the battery temperature. However, their speed is reduced when the vehicle is otherwise quiet (engine off, low speed, or stopped), presumably to prevent them from annoying the passengers.

A number of wires are added within the OEM HV battery box, including two HV cables that will exit the box and a short HV jumper cable through the HAL sensor. In addition, one or more low-voltage wires and taps may be added.[7]

1. Remove trunk carpet, all storage compartments, and spare tire.
2. **Be sure to remove the orange HV service plug.**
3. Install HV taps to the OEM NiMH battery pack.
   a. Remove the rear seat bottom, pull up the front edge, and unhook the rear edge.
   b. Remove the rear battery deck upholstery 2x tie down clips ?10 mm?, 2x tabs.
   c. Remove the driver's-side rear seat back 2 × 14 mm.
   d. Remove the driver's-side rear seat pillow upholstery 12 mm, pull up.
   e. Remove the driver's-side trunk panel upholstery 10 mm, 10 mm screw, tabs.
   f. Remove the driver's-side OEM battery brace 7 × 12 mm.
   g. Remove OEM battery ECU cover 2 × 10 mm, 2 × 10 mm nuts.
      - Optional steps. Caution: This will expose you to the OEM HV modules.
      - Remove OEM battery input air duct 2x tape, 1 sensor.
      - Remove passenger-side OEM battery brace 7 × 12 mm.
      - Remove OEM battery lid 4 × 12 mm, 4 × 10 mm, 4 × 10 mm nuts, 1 tab at end.
   h. Add new HV cables, notch here, trim there, etc., etc.

i. Reassemble in reverse order.

4. Replace orange HV service plug.

Mount PFC Charger

Install the PFC charger in the driver's-side rear wheel well.

1. Locate and drill mount points from the pattern.
2. Install ventilation housing and fans.
3. Mount the PFC charger.

Install Battery Box

1. Locate, drill, and press nut body/frame mount points.
2. Attach the electronics box to the bottom of the lower battery box frame.
3. Install and fasten the lower battery box frame.
4. Populate the box with batteries.
5. Install battery regulators and interconnects.
6. Install and fasten the upper battery box frame.

Interlock Box

The interlock box connects all the components of the kit. The low-power signals from the CAN-View relays and other OEM systems signals enter the interlock box. These signals are then passed on to the charger to enable full-power mode and two voltage settings; they can also enable full-speed mode for the OEM battery cooling fan; see Prius OEM battery fan control info. The high-power cables from the OEM

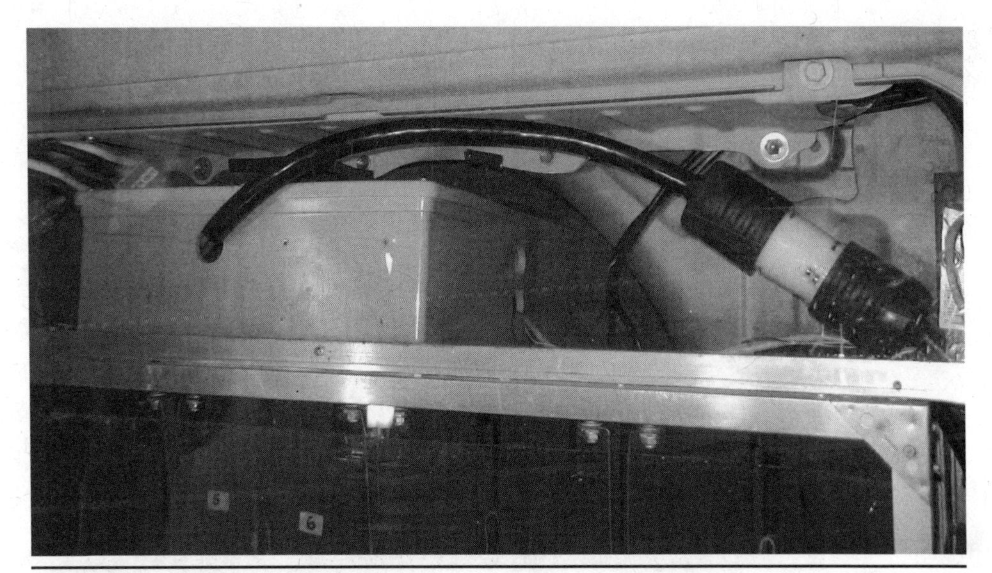

Figure 10-28 The PiPrius interlock box.

battery, grid charge port, charger input, and charger output connect here. The box contains the high-power contactors that switch from charge mode to PHEV mode. The PHEV battery pack interpack fuse or breaker and the AC line power circuit breaker are found in this box. The interlock box has been created as a 10" square plastic utility case, an aluminum box that fits the passenger-side cubby compartment, and an aluminum rectangular box mounted near the charger in the universal modular battery box.[10]

Figure 10-29 An early version (an early plastic box) of the PiPrius interlock box.

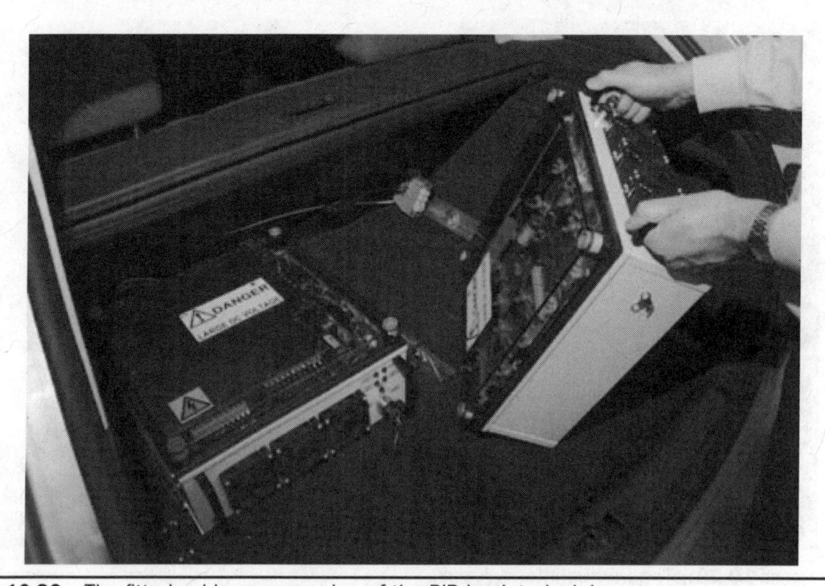

Figure 10-30 The fitted cubby area version of the PiPrius interlock box.

Figure 10-31 Placing the new Prius batteries into the trunk.

Figure 10-32 Installation of the battery box for the Ford Escape PHEV. Source: Elithion.

FIGURE 10-33 Installation of the battery box for the PRIUS+. Source: EAA-PHEV.

Figure 10-34 shows the modular version of the PiPrius battery box. The interlock box module can be seen on the far left of the four battery bank modules. Each battery bank contains six 12-V, ~20-Ah batteries and a battery regulator tray.

Figure 10-35 shows the modular version of the PiPrius battery box with the Lexan cover installed. The interlock box module is on the far left, and there are four battery bank modules. Each battery bank contains six 12-V, ~20-Ah batteries and a battery regulator tray.

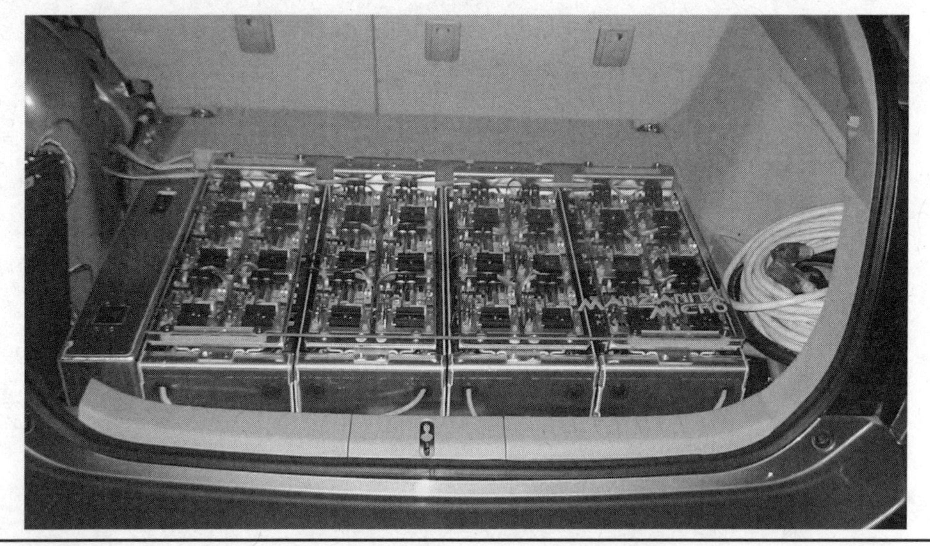

FIGURE 10-34 The modular version of the PiPrius battery box (far left).

Figure 10-35 The modular version with cover box installed.

Figure 10-36 Lithium-ion battery pack in CalCars' EnergyCS/EDrive converted Prius.

Regulators

The battery regulators monitor each 12-V battery to make sure that none of them are ever overcharged or overdischarged. During the charge cycle, each regulator is capable of dissipating x watts of power and communicating with the charger in order to lower the charge rate. During the discharge cycle, each regulator communicates with the charger to lower the *dis*charge rate once a battery reaches

FIGURE 10-37 Battery regulators for the PiPrius modular battery box.

FIGURE 10-38 Battery regulator tray for the PiPrius modular battery box.

about 10.5 V under load to prevent the batteries from being overdischarged. This regulation system should protect and prolong the life of the battery pack, and also help to diagnose any issues that may arise with weak batteries. Digital versions of the regulators are also capable of communications with a computer over a serial RS232 bus to report their current state and to accept new parameters.[11]

High-Current System

First you attach the high-current components, then you pull the AWG 2/0 cable to connect them (this is another step where scheduling an inside helper is appropriate). In Figure 10-39, notice that there are seven major components in the high-current line (in addition to batteries and charger—these we'll save for later):

- Series DC motor
- Motor controller
- Circuit breaker
- Main contactor
- Safety fuse
- Ammeter shunt(s)
- Safety interlock

FIGURE 10-39 High-current system to a lead-acid battery tie.

Figure 10-40 Inside the OEM battery in the back of the car (Toyota Prius).

Test Data Display

A variety of data are available through the instrument panel.

To start the test display:

- Start with the ignition off.
- Press and hold the Trip Reset button on the instrument panel.
- Turn the ignition to On (not Start).
- Wait until the display in the speedometer shows "test."
- Release the Trip Reset button.

Each time you press the Trip Reset button, a new set of data are displayed.

Paint, Polish, and Sign

After everything is running the way you like, it's time to "pimp your ride." Why? A car is an extension of oneself. The way the exterior of a vehicle looks reflects you; you are proud of your work, and you want to show it off in its best light. But using an outside "100 mpg" sign like the one Felix Kramer has on his car, as seen in

Table 10-11 Test Display Data

Press	Prefix	Values	Range	Description
0	test	—	—	Initial entry into test mode
1	gage	—	—	Test Sweep of all gauges from min to max
2	—	ALL	Blackout	Prove-out of all segments on odometer display
3	—	ALL	Blackout	Prove-out of all segments on message center display
4	bulb	—	—	Lights all bulbs/LEDs (look for "THEFT" bulb)
5	r	####		Returns all bulbs/LEDs to normal operation
6	nr	####		Hexadecimal code
7	EE	##		Hexadecimal code
8	dt	####		Hexadecimal code for manufacture date
9	CFI	##		Hexadecimal code
10	CF2	##		Hexadecimal code
11	CF3	##		Hexadecimal code
12	CF4	##		Hexadecimal code
13	CF5	##		Hexadecimal code
14	CF6	##		Hexadecimal code
15	DTC	nOnE		Diagnostic Trouble Code (You want this to say nOnE)
16	E	###.#	000.0–127.0	Speed in English to the tenth of a MPH
17	—	###.#	000.0–205.0	Speed in Metric to the tenth of a kmPH
18	t	####	0000–7000	Tachometer to nearest 1 RPM
19	F	###	000–255	Fuel IPHEVel analog/digital ratio input to instrument panel
20	FP	###	000–255	Fuel present IPHEVel status as an amount out of 255 = full
21	CA	##.#	00.0–40.0	Kilowatt value being used (+)/produced (–)
22	SOC 1	##	00–?	CAN message status to message center 00 = normal
23	ET	###	000–127	Engine Temperature in degrees Celsius (°C)
24	BT	###	000–127	HV Battery Temperature in degrees Celsius (°C)
25	ODO	###	000–255	Rolling count used to calculate odometry
26	TR	##.##	00.00–99.99	Trip odometer in miles and hundreths of a mile
27	NCS-	#		Message Center Status
28	BAT	##.#	00.0–19.9	Standard battery voltage reading
29	D	###	000/124 /255	Position of dimmer switch: 000 = up 124 = down 255 = off
30	RH5	##	00–21	Instrument cluster dimmer value: 00 = off 21 = max. bright
31	HLPS-	#	0–1	Status of parking / headlamps: 0 = off 1 = on
32	IIN-	#	0–1	Key in ignition: 0 = no 1 = yes

(continued)

Table 10-11 Test Display Data *(continued)*

Press	Prefix	Values	Range	Description
33	DOOR-	#	A or C	Driver door status: A = ajar C = closed
34	STBT-	#	0 or 6	Driver seatbelt status: 0 = buckled 6 = not buckled
35	PRND	##		Last value input to TRS from the PCM
36	PAR-	#	0 or 6	Status of park: 0 = in park 6 = not in park
37	CR-	#	0 or 6	Status of START: 0 = key in start 6 = key not in start
38	ACC-3	#	0 or 6	Status of ACC: 0 = key in ACC 6 = key not in ACC
39	Ch-	#		Chime: The chime that last sounded, or is currently sounding
40	ChE	##		2-bit MIL telltale data–Malfunction Indicator Lamp
41	OPS-	#	0 or 6	Oil Pressure Sensor: 0 = on 6 = off
42	TT1	##		Hexadecimal code
43	TT2	##		Hexadecimal code
44	TT3	##		Hexadecimal code
45	THFT	##	(14)	Anti-theft visual indicator mode. THEFT LIGHT STATUS
46	4b4	##		2-bit 4x4 message (if equipped)
47	361	##		Hexadecimal code
48	368	##		Hexadecimal code
49	3612	##		Hexadecimal code
50	369	##		Hexadecimal code
51	PA	##		Hexadecimal code
52	PADO	##		Hexadecimal code
53	PB	##		Hexadecimal code
54	PH	##		Hexadecimal code
55	PJ	##		Hexadecimal code
56	PL	##		Hexadecimal code
57	PCAN	##		Hexadecimal code
58	PT	##		Hexadecimal code
59	PUU	##		Hexadecimal code
60	BAT	###	000–255	8-bit value for standard battery voltage readings
61	AD2	###	000–255	8-bit value for dimmer readings
62	AD3	###	000–255	8-bit value for fuel IPHEVel readings
63	AD4	###	000–255	8-bit value for oil pressure (150–160 = normal with stock oil)
64	gage	—	—	Goes back to start and cycles through all features again

Chart courtesy of gpsman1@yahoo. Source: EAA-PHEV.

Chapter 3 (or in Figure 10-41), is the only way you can tell that the Prius or Ford Escape or *any* hybrid car you drive is part electric car.

FIGURE 10-41 Look at me—100 mpg!!! Source: EAA-PHEV.

FIGURE 10-42 Felix Kramer knows how to show it off. Source: EAA-PHEV/CalCars.

Because few people (if any) are likely to examine your PHEV conversion closely, you have to advertise instead. A couple of well-placed large letters is all it takes—or a plug-in hybrid electric vehicle license tag.

Put Yourself in the Picture

Now that you've seen how it's done, you can do it yourself. It's a very simple project that virtually anyone can accomplish today, just by asking for help in a few appropriate places and taking advantage of today's kit components and prebuilt conversion packages. Imagine yourself in Figures 10-43 through 10-47 and take the steps to make it happen.

FIGURE 10-43 Toyota Prius with EV-mode button.

FIGURE 10-44 Plug-In Center crew with their Prius PHEV.

Figure 10-45 HybridsPlus battery pack installation.

Figure 10-46 Battery pack and BMS installation in back of Ford Escape PHEV. Source: Elithion.

Figure 10-47 What else is there to do but jump for joy! Source: Elithion.

CHAPTER 11

Maximize Your Plug-In Hybrid Electric Vehicle Enjoyment

Now that you are driving for only pennies a day,
it takes little more to make your pleasure complete.

Once you've driven your PHEV conversion around the block for the first time, it's time for you to start planning for the future. Since the vehicle will have been registered as a hybrid, you need to license and insure it so that you can drive it farther than just around the block. You also need to learn how care for it so that you maximize your driving pleasure and its economy and longevity.

Licensing and Insurance Overview

Your vehicle goes 75 mph, and Ford has already guaranteed that its chassis complies with FMVSS and NHTSA safety standards.

Which end of the spectrum your PHEV conversion of today resembles directly determines its ability to clear any licensing and insuring hurdles. Let's take a closer look at each area.

Getting Licensed

The licensing of any vehicle is under the jurisdiction of the state in which it resides. Although state motor vehicle codes are based on common federal standards, each of them is just a little bit different. While the chassis of your internal combustion engine vehicle conversion may be fully compliant with federal FMVSS and NHTSA safety standards, you need to check out what your state's motor vehicle code says. If you're doing a from-the-ground-up PHEV, you'd be well advised to check out these rules and regulations in advance.

In general, few states have specific regulations for PHEVs. You'll find that states with larger vehicle populations, such as California, New York, and Florida, are on the leading edge in terms of establishing guidelines for PHEVs. Check with your own state's motor vehicle department to be sure.

As for the licensing process, most of the people who work for the Department of Motor Vehicles and/or the Department of Environmental Protection in each state are far more involved with smog certification or DEQ (Department of Environmental Quality) regulations. It is very important that you check the rules and regulations in your own state to see what the process is for allowing a converted hybrid electric vehicle to receive license plates.

In most states, you can receive a tax credit for an electric vehicle, and there is also a federal tax credit for electric vehicles. In some areas, you may be entitled to a reduction in your electric power rate. Check with your local utility and with your city and state governments to see if you and your PHEV are entitled to something similar in your area.

Getting Insured

Insurance is roughly the same as licensing. You're not likely to have any trouble with your PHEV if it's been converted from an internal combustion engine chassis. While explaining to your insurance carrier that the vehicle has been converted to electric power may be quite a process, having a vehicle identification number (VIN) and approval from the State Department of Motor Vehicles will make most large national insurance companies feel at ease with underwriting insurance coverage for a PHEV. Verify that your design meets insurance requirements in advance.

Safety Footnote

My basic assumption in this section is that you have put safety high on your list of desirable characteristics for your converted PHEV. This line of reasoning assumes that you have left the safety systems of the original internal combustion engine chassis intact: lights, horn, steering, brakes, parking brake, seat belts, windshield wipers, and so on. It also assumes that you are thinking "safety first" when installing your new PHEV components.

Driving and Maintenance Overview

A PHEV is easier to drive and requires less maintenance than its internal combustion counterpart. But because its driving and maintenance requirements are different from those of an internal combustion vehicle, you'll need to adjust the driving habits you acquired when driving an internal combustion engine vehicle. The driving part is very similar to the experience of a lifelong stick-shift driver who drives an automatic-transmission vehicle for the first time.

Driving Your PHEV

Your PHEV conversion may still look like its internal combustion engine ancestor, but it drives very differently. Here's a short list of reminders.

Economical Driving

If you keep an eye on your ammeter while you're driving, you'll soon learn the most economical way to drive, shift gears, and brake. For maximum range, the objective is to use the least current at all times. You'll immediately notice the difference when you're drag racing or going up hills—either alter your driving habits or plan on recharging more frequently.

Coasting

If you don't have regeneration, coasting in a PHEV is unlike anything you've encountered in your internal combustion engine vehicle—there's no engine compression to slow you down. You need to learn how to "pulse" the engine correctly and when. I have ridden with drivers who floor the accelerator for three seconds, then coast, floor it, then coast, and so on. Heavy pulsing is not good for the vehicle, and it wastes energy. In most driving, a steady foot is better. Light pulsing is an advantage only when little power is needed.

Regeneration

Regenerative braking is a mechanism that reduces the speed of the vehicle by converting some of its kinetic energy into another useful form of energy. This captured energy is then stored for future use or fed back into a power system for use by other vehicles. For example, electric regenerative brakes in electric railway vehicles feed the generated electricity back into the supply system. In battery electric and hybrid electric vehicles, the energy is stored in a battery or a bank of capacitors for later use. Other forms of energy storage that may be used include compressed air and flywheels.[1]

Many people have now experienced coasting in a hybrid electric car. Like any hybrid, and by definition, a PHEV is designed to be as frictionless as possible, so take advantage of this great characteristic. Learn to pulse your accelerator and coast to the next light or to the vehicle ahead of you in traffic.

Hybrid electric car owners (especially New York City taxi owners) understand this concept. When you accelerate, you do not need to floor the accelerator and then coast. You can step on the accelerator slowly, then take your foot off it and coast. When you coast, you are using the regenerative braking. This is much smoother and a more efficient use of the vehicle, since it is moving forward and charging at the same time. While there are plenty of people who own converted PHEVs who like to floor it and coast, regeneration can be a very big help in most stop-and-go driving.

The PHEV conversion has no effect on how much energy can be recovered from braking during normal driving. Yet regenerative braking is not as impressive in a PHEV as it is in other hybrid electric vehicles: because the PHEV battery can hold so much more energy than the standard HEV battery, the energy from regenerative braking during normal driving is a small portion of the energy in the pack. However, when you are coming down a long, steep hill, the battery in a PHEV may be able to keep on taking energy from regenerative braking for a long time (about 1-1/2 hours), unlike the battery of a standard HEV, which is quickly filled.[2]

Caring for Your PHEV

Now that you are driving for only pennies a day, it takes little more to make your pleasure complete. Actually, a properly designed and built PHEV conversion requires surprisingly little attention compared to an internal combustion engine vehicle. It comes down to the care and feeding of your batteries, minimizing friction, and preventive maintenance.

Battery Care

Of course, you are going to be charging your batteries on a regular basis, using the guidelines in Chapter 8, so battery maintenance really comes down to checking periodically to see that your batteries are properly watered. On a Prius (for example), additional diagnostics can be viewed on the public DigiKey parts list for the LED board. Figure 11-1 shows the PRIUS+ control board schematic, and Figure 11-2 shows the LED board schematics from the latest PRIUS+ conversion.

Lubricants

The weight or viscosity of your drivetrain fluids (transmission and rear axle lubricant) also contributes to losses on an ongoing basis, so experiment with lightweight lubricant in both these areas. The PHEV conversion puts a much smaller design load on the mechanical drivetrain, so you ought to be able to drop down to a 50-weight lubricant for the rear axle and a lower-loss transmission fluid grade. Consider low-loss synthetic lubricants.

Checking Connections

Preventive maintenance mostly involves checking the high-current wiring connections for tightness periodically. Use your hands here. Warmth is bad—it means a loose connection—and anything that moves when you pull it is also bad. A few open-end and box wrenches ought to make quick work of your retightening preventive maintenance routine.

Emergency Kit

While carrying extra onboard weight is a no-no, you ought to carry a small highway kit to give you on-the-road peace of mind, knowing you have planned for most contingencies. At a minimum, your kit should have a small fire extinguisher, a small bottle of baking soda solution, a small toolkit (wrenches, flat and wire-cutting pliers, screwdrivers, wire, and tape), and a heavy-duty charger extension cord with multiple adapter plugs (male and female).

FIGURE 11-1 PRIUS+ control board schematic. Source: www.eaa-phev.org/images/b/b2/EAA-PHEV-PRIUS-ControlBdSchematic.png.

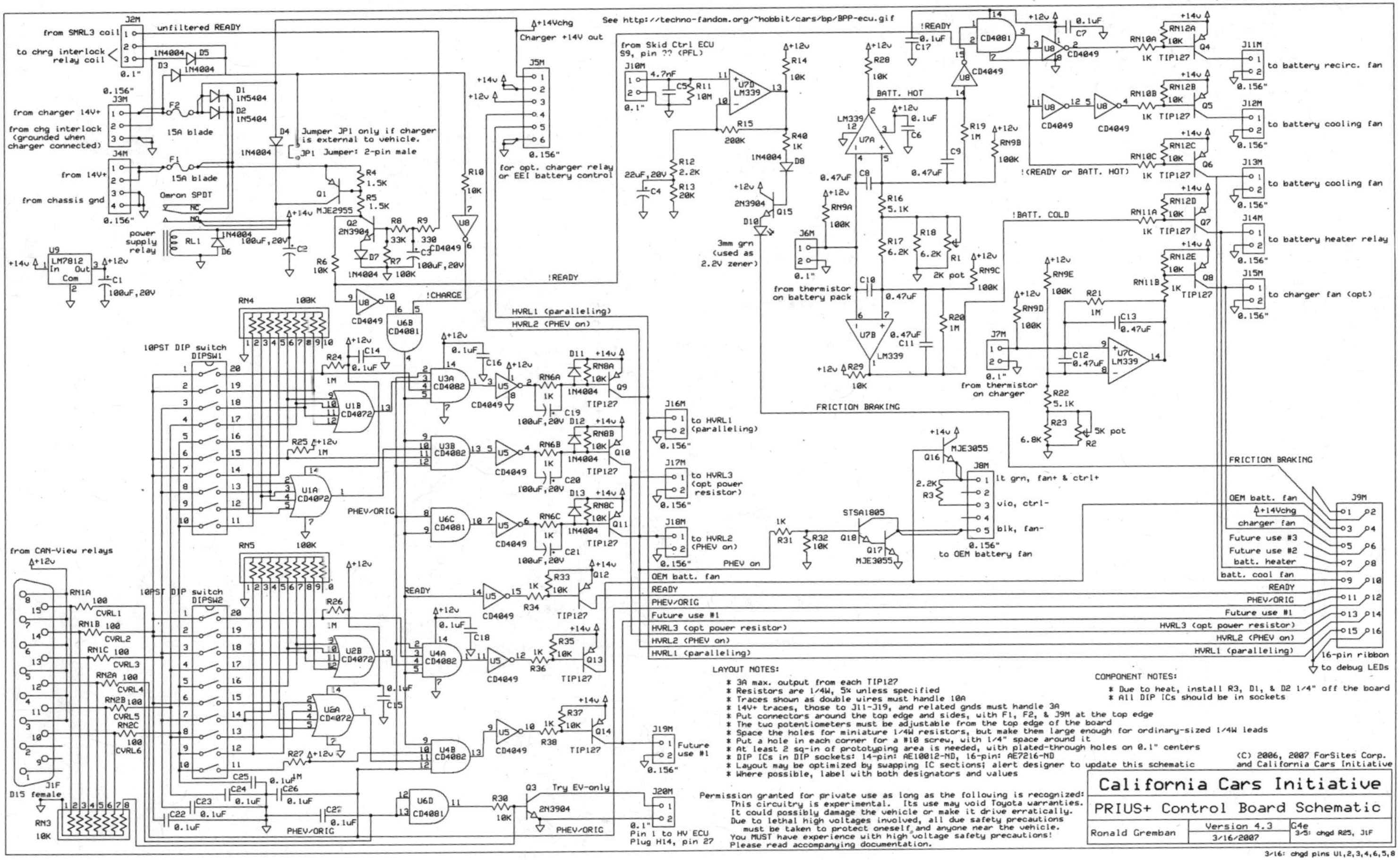
from SMRL3 coil
to chrg interlock relay coil
unfiltered READY
from charger 14V+
from chg interlock (grounded when charger connected)
15A blade
from 14V+
from chassis gnd
Omron SPDT
power supply relay
Jumper JP1 only if charger is external to vehicle.
JP1 Jumper: 2-pin male
Charger +14V out
for opt. charger relay or EEI battery control
See http://techno-fandom.org/~hobbit/cars/bp/BPP-ecu.gif
from Skid Ctrl ECU S9, pin ?? (PFL)
3mm grn (used as 2.2V zener)
from thermistor on battery pack
from thermistor on charger
BATT. HOT
!(READY or BATT. HOT)
!BATT. COLD
FRICTION BRAKING
to battery recirc. fan
to battery cooling fan
to battery cooling fan
to battery heater relay
to charger fan (opt)
from CAN-View relays
D15 female
10PST DIP switch
to HVRL1 (paralleling)
to HVRL3 (opt power resistor)
to HVRL2 (PHEV on)
Future use #1
Pin 1 to HV ECU Plug H14, pin 27
Try EV-only
lt grn, fan+ & ctrl+
vio, ctrl-
blk, fan-
to OEM battery fan
16-pin ribbon to debug LEDs
LAYOUT NOTES:
* 3A max. output from each TIP127
* Resistors are 1/4W, 5% unless specified
* Traces shown as double wires must handle 10A
* 14V+ traces, those to J11-J19, and related gnds must handle 3A
* Put connectors around the top edge and sides, with F1, F2, & J9M at the top edge
* The two potentiometers must be adjustable from the top edge of the board
* Space the holes for miniature 1/4W resistors, but make them large enough for ordinary-sized 1/4W leads
* Put a hole in each corner for a #10 screw, with 1/4" space around it
* At least 2 sq-in of prototyping area is needed, with plated-through holes on 0.1" centers
* DIP ICs in DIP sockets: 14-pin: AE10012-ND, 16-pin: AE7216-ND
* Layout may be optimized by swapping IC sections; alert designer to update this schematic
* Where possible, label with both designators and values
COMPONENT NOTES:
* Due to heat, install R3, D1, & D2 1/4" off the board
* All DIP ICs should be in sockets
Permission granted for private use as long as the following is recognized:
This circuitry is experimental. Its use may void Toyota warranties.
It could possibly damage the vehicle or make it drive erratically.
Due to lethal high voltages involved, all due safety precautions must be taken to protect oneself and anyone near the vehicle.
You MUST have experience with high voltage safety precautions!
Please read accompanying documentation.
(C) 2006, 2007 ForSites Corp. and California Cars Initiative
California Cars Initiative
PRIUS+ Control Board Schematic
Ronald Gremban
Version 4.3
3/16/2007
G4e
3/5: chgd R25, J1F
3/16: chgd pins U1,2,3,4,6,5,8

Figure 11-2 LED board schematic from the latest PRIUS+ conversion.

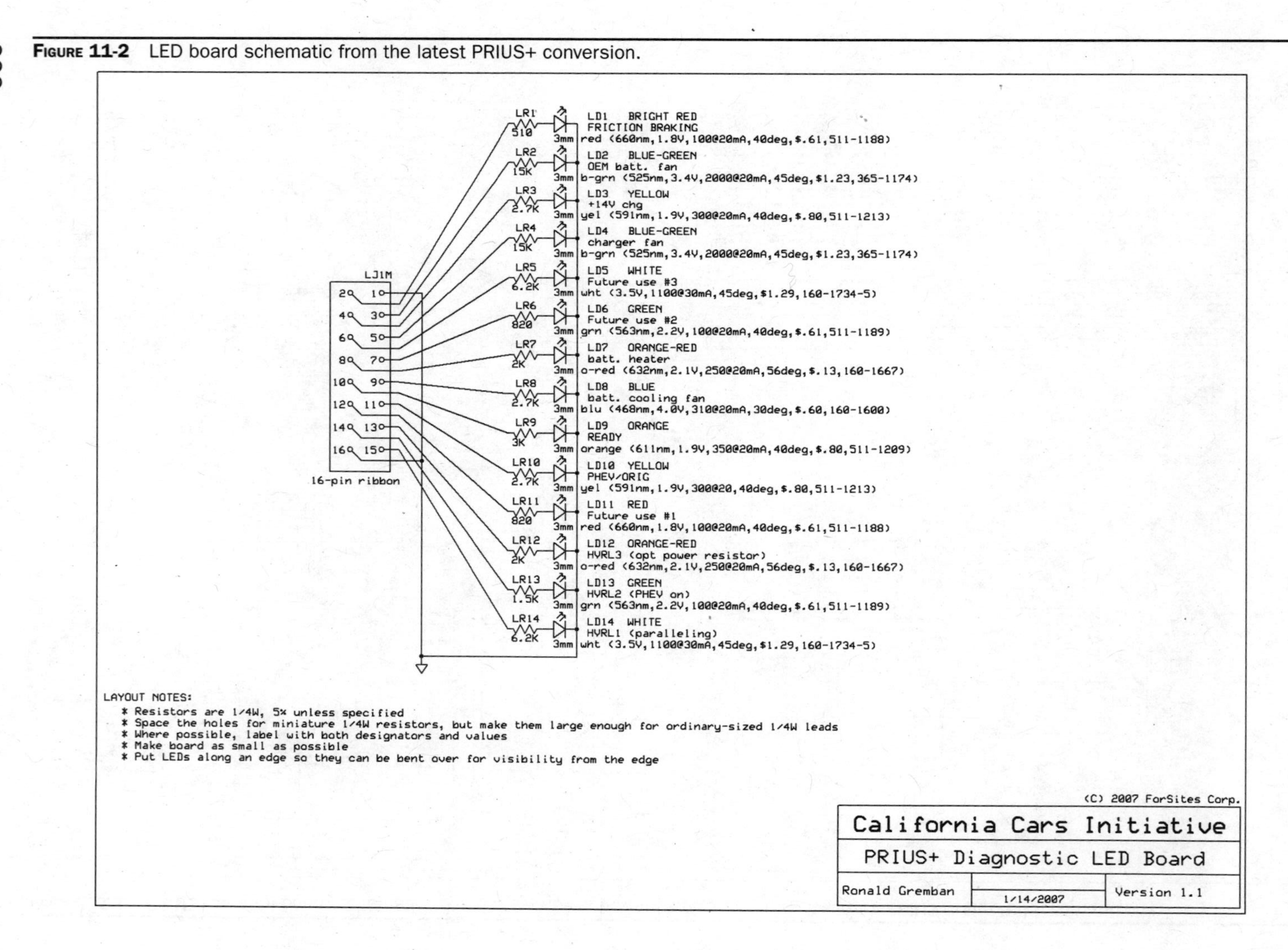

Will This Conversion Affect My Warranty?

As for *warranties*, there is legal precedent for the conclusion that original auto warranties cannot be voided completely by modifications—only the part(s) affected by the retrofit will have their warranties affected. Typically, hybrid cars have four or five separate warranties. Car companies could decide that conversions void some or all of the hybrid system warranty, but they may fear that doing so will tarnish their green image. Drivers of converted cars have received high levels of cooperation from dealer service departments so far. We agree with Toyota's comments, which is why we advocate mass production of PHEVs:

> We're immensely gratified that some enthusiasts, in a tacit endorsement of the hybrid concept, are, on a vehicle-by-vehicle basis, converting Toyota hybrids to plug-in technology. But doing one-by-one conversions is a different kettle of volts from making this technology viable for the sale of hundreds of thousands of cars, at an affordable price, with a reasonable reliability expectations and reasonable warranty, serviceable at any Toyota dealer anywhere.[3]

By law, the car manufacturer cannot void your warranty because of modifications to the vehicle that are not related to the area that needs service. Toyota shouldn't void your engine warranty if you install a cup holder. Similarly, installing a different battery pack shouldn't void your warranty for anything but the battery pack itself and the high-voltage components in the car.

The Specialty Equipment Market Association (SEMA) discusses the issue as well. Here's an excerpt:

1. **The Magnuson-Moss Warranty Act (15 U.S.C. 2302(C))** This federal law regulates warranties for the protection of consumers. The essence of the law concerning aftermarket auto parts is that a vehicle manufacturer may not condition a written or implied warranty on the consumers using parts or services which are identified by brand, trade, or corporate name (such as the vehicle maker's brand) unless the parts or service are provided free of charge. The law means that the use of an aftermarket part alone is not cause for denying the warranty. However, the law's protection does not extend to aftermarket parts in situations where such parts actually caused the damage being claimed under the warranty. Further, consumers are advised to be aware of any specific terms or conditions stated in the warranty which may result in its being voided.

See also the Federal Trade Commission's description of the Magnuson-Moss Act.

This comes from a Toyota representative:

> Any modifications that are made to the vehicle will affect the warranty for the Toyota parts that are associated to the modification. The entire vehicle warranty would not be

voided. If one of the components the customer installs causes damage to a Toyota part on the vehicle that part would no longer be under warranty. Toyota Fleet.[4]

Cell Board Troubleshooting: Diagnostics and Repair of Cell Boards for Standard Battery Management Systems

There are four ways of troubleshooting a cell board:

- Observing its behavior in the system
- Testing it without a test fixture
- Testing it with a test fixture
- Replacing it with a known good cell board

Observing Behavior

- With the battery management system (BMS) controller off, note whether the LED stays off.
  - If the LED stays on, the cell board's load is permanently on; replace the cell board.
- With the BMS controller on, note whether the LED blinks once a second.
  - If the LED stays on, and it's the only one that does so, the cell board is not measuring correctly; replace the cell board.
  - If the LED doesn't blink, and it's the only one that doesn't, the cell board is not alive; replace the cell board.
  - If the LED doesn't blink, nor do the other ones in that bank, but the cell boards in other banks do blink, check the cell board on the most positive cell or the cable to it from the BMS controller.

Testing without a Test Fixture

- Make sure that either the BMS controller is not powered or the cell board is not connected to the BMS controller.
- Measure the cell voltage.
  - If the cell voltage is above 2.5 V, this part is OK.
  - If the cell voltage is below 2.5 V, either the cell has been overdischarged or the cell board is bad and it discharged the cell.
- Measure the voltage at the C+ terminal with respect to the B+ terminal.
  - If the voltage is the same as the cell voltage, this part is OK.
  - If the voltage is 0 V, there is a short between the C+ and the B+ terminals.
  - If the voltage is neither 0 nor equal to the cell voltage, the problem is in the cell board on a cell more positive than this one.
- Disconnect one end of the cell board from the cell.

- Reconnect it.
- Normally, the LED blinks twice and repeats that a few times, then goes off.
  - If the LED blinks twice and repeats that a few times, then goes off, this part is OK.
  - If the LED stays on, the cell board's load is permanently on; replace the cell board.
  - If the LED doesn't come on, the cell board is dead; replace it.

Testing with a Test Fixture

This method is for loose cell boards (those that are not installed on cells). You will need to build a relatively simple cell board test fixture. Testing must be done at room temperature, or "too cold" or "too hot" failures will result.

- Power up the test fixture.
- Connect the cell board to the test fixture (if there are multiple sections, connect just one of the sections).
- Press the test fixture's pushbutton.
- Note the behavior of the LED on the cell board, and use Table 11-1 to troubleshoot.

For cable-mounted sensors, inversion may be needed, depending on which way the sensor is mounted.

Replacement

- Disconnect the communication wires to the adjacent cell boards (if applicable).
- Disconnect the communication cables to the BMS controller (if applicable).
- Disconnect the cell board.
- Install a known good cell board.
- Reconnect the communication wires and/or the communication cables, as appropriate.
- Test the system.

Table 11-1 Troubleshooting Notes on Cell Boards

LED	Result	Problem	Cause	Possible Components
1 blink	Pass			
2 blinks	Fail	Input pulse too short	No feedback from C– to C+	Micro pin 3 shorted, Rbus open, Qout bad/bridged, test fixture C– pin
3 blinks	Fail	Input pulse doesn't end	Unable to ground ouput	Micro pin 3 open, Qout bad/bridged, C+ shorted to B+, C– shorted to B–
4 blinks	Fail	Temperature = 0	No Therm reading	Thermistor shorted Thermistor's resistor open/missing Micro pin 7 and 8 bridged
5 blinks	Fail	Temperature too low	Therm reading is low	Too hot, bad thermistor, Rtherm value too high
6 blinks	Fail	Temperature too high	Therm reading is high	Too cold, bad thermistor, Rtherm value too low
7 blinks	Fail	Temperature = FFh	No Thermistor	Thermistor missing/open, open trace
8 blinks	Fail	Ref reading = 0	No Vref input	Uref bad/bridged, R4 open/missing, open trace
9 blinks	Fail	Ref reading too low	Vref reading is low	Micro pin 5 and 6 bridged, open trace
10 blinks	Fail	Ref reading too high	Vref reading is high	MOSFET shorted/bridged, open trace
11 blinks	Fail	Ref reading = FFh	No Vref	Uref missing/open/bad, open trace
12 blinks	Fail	No load	Load doesn't come on	MOSFET missing/open/bad, Rload open, open trace
Off	Fail	No input pulse	Doesn't see fixture	Not programmed, bad board, bad LED, Rin open, Cin shorted, pin 4 open, trace, test fixture C+ pin
Always on	Fail	Load on	Load drive always on	MOSFET shorted/bridged

CHAPTER 12
Sources

Knowledge is but the beginning of wisdom.
—Darwin Gross, *Universal Key*

One of the most valuable benefits this book provides is information about where to go and who to see. While there are many sources, you'll find your needs best served by confining your initial search to just a handful until you get acquainted with the EV field. Please note that you must do your own reading, searching, listening, and talking before shopping to buy, convert, or build in order to get the best products and price.

The best guides for individuals interested in EVs are usually found in, around, and through the Electric Auto Association.

Less Is More

I have chosen to give you a few sources in each category to get you started rather than attempting to list everything and get you confused. The rest of this chapter is divided into four sections:

- Clubs, associations, and organizations
- Manufacturers, converters, and consultants
- Suppliers
- Books, articles, and literature

Clubs, Associations, and Organizations

The original Electric Auto Association, whose logo appears in Figure 12-1, has numerous local chapters. There are offshoots from the original, and new local

FIGURE 12-1 The Electric Auto Association's logo says it all. Registered trademark of the Electric Auto Association.

entities having no connection with the original. There are also associations and organizations designed to serve corporate and commercial interests rather than individuals. Each one of these has its own meetings, events, and newsletter.

PHEV Organizations

California Cars Initiative
P.O. Box 61045
Palo Alto, CA 94306
Phone: (650) 520-5555
www.calcars.org
Contact: Felix Kramer <info@calcars.org>

Electric Auto Association
www.eaaev.org
Founded in 1967, this is the oldest, largest organization, and has consistently been the best source of EV information for the individual. The newsletter subscription is well worth the price of the membership dues. Recent newletters have averaged 16 to 20 pages and provide information on current EV news and happenings. I can't emphasize enough the invaluable knowledge on tap in the members of this organization. The experience level of the various drivers who have built their own conversions can shorten one's learning curve substantially. Information on and photos of the many samples of conversions done over the years by this organization can be found at www.evalbum.com.

Electric Drive Transportation Association
1101 Vermont Avenue, NW, Suite 401
Washington, DC 20005
(202) 408-0774
www.electricdrive.org
EDTA is the preeminent industry association dedicated to advancing electric drive as a core technology on the road to sustainable mobility. As an advocate for the

adoption of electric drive technologies, EDTA serves as the unified voice for the industry and is the primary source of information and education related to electric drive. Its membership includes a diverse representation of vehicle and equipment manufacturers, energy providers, component suppliers, and end users.

National Electric Drag Racing Association
3200 Dutton Avenue, #220
Santa Rosa, CA 95407
www.nedra.com
The National Electric Drag Racing Association (NEDRA) exists to increase public awareness of electric vehicle performance and to encourage, through competition, advances in electric vehicle technology. NEDRA achieves this by organizing and sanctioning safe, silent, and exciting electric vehicle drag racing events.

Plug In America
6261 East Fox Glen
Anaheim, CA 92807
Phone: (714) 974-5647
www.pluginamerica.com
Contact: Linda Nicholes <Linda@pluginamerica.com>
Meetings: Please contact for details

Plug-In Partners
c/o Partnership Services Coordinator
721 Barton Springs Road
Austin, TX 78704
Phone: (512) 322-6511
Fax: (512) 322-6037
Deputy Coordinator
Austan Librach, P.E., AICP

Set America Free
7811 Montrose Road, Suite 505
Potomac, MD 20854
www.setamericafree.org/home.html

Solar and Electric Racing Association
11811 N. Tatum Boulevard, Suite 301
Phoenix, AZ 85028
Phone: (602) 953-6672
Organizes annual Solar and Electric 500 in Phoenix and promotes electric vehicles.

Here are some EAA local chapters—maybe there's one near you.

Canada

Durham Electric Vehicle Association
P.O. Box 212
Whitby, ON L1N 5S1
Phone: (905) 706-6647
www.durhamelectricvehicles.com
Contact: JP Fernback <mail@durhamelectricvehicles.com>
Meetings: First Thursday of the month from September until June

Electric Vehicle Council of Ottawa
P.O. Box 4044
Ottawa, ON K1S5B1
Phone: (613) 271-0940
www.evco.ca
Contact: Alan Poulsen <info@evco.ca>
Meetings: 7:30 pm–10:00 pm, last Monday of the month

Electric Vehicle Society of Canada–Toronto
88 Lake Promenade
Etobicoke, ON M8W-1A3
Phone: (416) 255-9723
www.evasociety.ca
Contact: Neil Gover <neil@ontarioev.ca>
Meetings: 7:30 pm, third Thursday of the month (except July and August)

Vancouver Electric Vehicle Association
4053 West 32nd Avenue
Vancouver BC V65 1Z5
Phone: (604) 527-4288
www.veva.bc.ca
Contact: Haakon MacCallum <info@veva.bc.ca>
Meetings: 7:30 pm, third Wednesday of the month (please check website for details)

Alaska

Alaska EAA
2550 Denali, Suite 1
Anchorage, AK 99503
www.alaskaEVA.org
Phone: (907) 868-5710
Contact: Mike Willmon <electrabishi@ak.net>
Meetings: 8:00 pm–9:00 pm, third Friday of the month

Arizona

Flagstaff EAA

6215 Rinker Circle
Flagstaff, AZ 86004
Phone: (928) 637-4444
Contact: Barkley Coggin <cbcoggin@yahoo.com>
Meetings: 7:00 pm–9:00 pm, first Wednesday of the month

Phoenix EAA

1070 E. Jupiter Place
Chandler, AZ 85225
Phone: (480) 659-5513
www.phoenixeaa.com
Contact: Jim Stack <jstackeaa@yahoo.com>
Meetings: 9:00 am, fourth Saturday of the month

Tucson EVA II

4207 N. Limberlost Place
Tucson, AZ 85705
Phone: (520) 293-3500
www.teva2.com
Contact: John Barnes <johnjab@cox.net>
Meetings: 9:00 am, second Saturday of the month

California

Central Coast EAA

323 Los Altos Drive
Aptos, CA 95003
www.eaacc.org
Phone: (831) 688-8669
Contact: Will Beckett <will@becketts.ws>
Meetings: Call or see website for meeting information.

Chico EAA

39 Lakewood Way
Chico, CA 95926
Phone: (530) 899-1835
www.geocities.com/chicoeaa
Contact: Chuck Alldrin <chicoeaa@sunset.net>
Meetings: 11:00 am–1:00 pm, second Saturday of the month

East (SF) Bay EAA
2 Smith Court
Alameda, CA 94502-7786
Phone: (510) 864-0662
www.ebeaa.org
Contact: Ed Thorpe <EAA-contact@excite.com>
Meetings: 10:00 am–12:00 noon, fourth Saturday of the month

Education Chapter: San Diego State University, College of Engineering
6161 El Cajon Boulevard
San Diego, CA 92115
Phone: (619) 933-6058
Contact: James S. Burns, PhD <jburns@mail.sdsu.edu>
Meetings: Fourth Tuesday of each month during the year, except for December

EVA of San Diego
1638 Minden Drive
San Diego CA 92111
Phone: (858) 268-1759
www.evaosd.com
Contact: Bill Hammons <ncsdca@att.net>
Meetings: 7:00 pm, fourth Tuesday of the month

EVA of Southern California
35 Maracay
San Clemente, CA 92672
Phone: (949) 492-8115
Contact: Leo Galcher <leo4marg@mac.com>
Meetings: 10:00 am, third Saturday of the month

Greater Sacramento EAA
8392 West Granite Drive
Granite Bay, CA 95746
Phone: (916) 791-1902
Contact: Tim Hastrup <tim.hastrup@surewest.net>
Meetings: 12:00 noon, third Tuesday of February, May, August, and November

Konocti EAA
800 S. Main Street
Lakeport, CA 95453
Phone: (707) 263-3030
www.konoctieaa.org
Contact: Dr. Randy Sun <rsun@mchsi.com>
Meetings: 11:00 am, last Friday of the month

North (SF) Bay EAA
c/o Agilent Technologies
1400 Fountaingrove Parkway
Santa Rosa, CA 95403
Phone: (707) 577-2391 (weekdays)
www.nbeaa.org
Contact: Chris Jones <chris_b_jones@prodigy.net>
Meetings: 10:00 am–12:00 noon, second Saturday of the month, check website for details

San Francisco Electric Vehicle Association
1484 16th Avenue
San Francisco, CA 94122-3510
Phone: (415) 681-7716
www.sfeva.org
Contact: Sherry Boschert <info2007@sfeva.com>
Meetings: 11:00 am–1:00 pm, first Saturday of the month

San Francisco Peninsula EAA
160 Ramona Avenue
South San Francisco, CA 94080-5936
Phone: (650) 589-2491
Contact: Bill Carroll <billceaa@yahoo.com>
Meetings: 10:00 am, first Saturday of the month

San Jose EAA
20157 Las Ondas
San Jose, CA 95014
Phone: (408) 446-9357
www.geocities.com/sjeaa
Contact: Terry Wilson <historian@eaaev.org>
Meetings: 10:00 am, second Saturday of the month

Silicon Valley EAA
1691 Berna Street
Santa Clara, CA 95050
Phone: (408) 464-0711
www.eaasv.org
Contact: Jerry Pohorsky <JerryP819@aol.com>
Meetings: Third Saturday (January–November)

Ventura County EAA
283 Bethany Court, Thousand Oaks, CA 91360-2013
Phone: (805) 495-1026
Contact: Bruce Tucker <tuckerb2@adelphia.net>
www.geocities.com/vceaa
Meetings: Please contact Bruce for time and location

Colorado

Denver Electric Vehicle Council
6378 S. Broadway
Boulder, CO 80127
Phone: (303) 544-0025
Contact: Graham Hill <ghill@21wheels.com>
Meetings: Third Saturday monthly, contact George for time and location

Florida

Florida EAA
8343 Blue Cypress
Lake Worth, FL 33467
Phone: (561) 543-9223
www.floridaeaa.org
Contact: Shawn Waggoner <shawn@suncoast.net>
Meetings: 9:30 am, second Saturday of the month

Georgia

EV Club of the South
750 West Sandtown Road
Marietta, GA 30064
Phone: (678) 797-5574
www.evclubsouth.org
Contact: Stephen Taylor <sparrow262@yahoo.com>
Meetings: 6:00 pm, first Wednesday every even numbered month

Illinois

Fox Valley EAA
P.O. Box 214
Wheaton, IL 60189-0214
Phone: (630) 260-0424
www.fveaa.org
Contact: Ted Lowe <ted.lowe@fveaa.org>
Meetings: 7:30 pm, third Friday of the month

Kansas/Missouri

Mid America EAA

1700 East 80th Street
Kansas City, MO 64131-2361
Phone: (816) 822-8079
www.maeaa.org
Contact: Mike Chancey <eaa@maeaa.org>
Meetings: 1:30 pm, second Saturday of the month

Massachusetts

New England EAA

29 Lovers Lane
Killingworth, CT 06419
Phone: (203) 530-4942
www.neeaa.org
Contact: Bob Rice <bobrice@snet.net>
Meetings: 2:00 pm–5:00 pm, second Saturday of the month

Pioneer Valley EAA

P.O. Box 153
Amherst, MA 01004-0153
www.pveaa.org
Contact: Karen Jones <PVEAA@comcast.com>
Meetings: 2:00 pm, third Saturday of the month (January–June; September–November)

Minnesota

Minnesota EAA

4000 Overlook Drive
Bloomington, MN 55437
Phone: (612) 414-1736
www.mn.eaaev.org
Contact: Craig Mueller <craig.mueller@nwa.com>
Meetings: 7:00 pm–8:30pm CDT

Nevada

Alternative Transportation Club, EAA

2805 W. Pinenut Court
Reno, NV 89509
Phone: (775) 826-4514
www.electricnevada.org
Contact: Bob Tregilus <lakeport104@yahoo.com>
Meetings: 6:00 pm, monthly, see website or call for details

Las Vegas Electric Vehicle Association
2816 El Campo Grande Avenue
North Las Vegas, NV 89031-1176
Phone: (702) 636-0304
www.lveva.org
Contact: William Kuehl <bill2k2000@yahoo.com>
Meetings: 10:00 am–12:00 noon, third Saturday of the month

North Carolina

Coastal Carolinas Wilmington
1317 Middle Sound
Wilmington, NC 28411
Phone: (910) 686-9129
Contact: Page Paterson <pagepaterson@mac.com>
Meetings: Please contact for time and date

Electric Cars of Roanoke Valley
567 Miller Trail
Jackson, NC 27845
Phone: (252) 534-1258
Contact: Harold Miller <EV@schoollink.net>
Meetings: Please contact for time and date

Piedmont Carolina Electric Vehicle Association
1021 Timber Wood Court
Matthews, NC 28105
Phone: (704) 849-9648
www.opecthis.info
Contact: Todd W. Garner <tgarnercgarner@yahoo.com>
Meetings: Please contact for time and date

Triad Electric Vehicle Association
2053 Willow Spring Lane
Burlington, NC 27215
Phone: (336) 213-5225
www.localaction.biz/TEVA
Contact: Jack Martin <jmartin@hotmail.com>
Meetings: 9:00 am, first Saturday of the month

Triangle EAA
9 Sedley Place
Durham, NC 27705-2191
Phone: (919) 477-9697
www.rtpnet.org/teaa
Contact: Peter Eckhoff <teaa@rtpnet.org>
Meetings: Third Saturday of the month

Oregon

Oregon Electric Vehicle Association
19100 SW Vista Street
Aloha, OR 97006
www.oeva.org
Contact: Rick Barnes <barnes.rick@verizon.net>
Meetings: 7:30 pm, second Thursday of the month

Pennsylvania

Eastern Electric Vehicle Club
P.O. Box 134
Valley Forge, PA 19482-0134
Phone: (610) 828-7630
www.eevc.info
Contact: Peter G. Cleaveland <easternev@aol.com>
Meetings: 7:00 p.m., second Wednesday of the month

Texas

Alamo City EAA
9211 Autumn Bran
San Antonio, TX 78254
Phone: (210) 389-2339
www.aceaa.org
Contact: Alfonzo Ranjel <acranjel@sbcglobal.net>
Meetings: 3:00 pm CST, third Sunday of the month

AustinEV: the Austin Area EAA
P.O. Box 49153
Austin, TX 78765
Phone: (512) 453-2890
www.austinev.org
Contact: Aaron Choate <austinev-info@austinev.org>
Meetings: Please see website

Houston EAA
8541 Hatton Street
Houston, TX 77025-3807
Phone: (713) 218-6785
www.heaa.org
Contact: Dale Brooks <brooksdale@usa.net>
Meetings: 6:30 pm, third Thursday of the month

North Texas EAA
1128 Rock Creek Drive
Garland, TX 75040
Phone: (214) 703-5975
www.nteaa.org
Contact: John L. Brecher <jlbrecher@verizon.net>
Meetings: Second Saturday of the month

Utah

Utah EV Coalition
325 E. 2550 N #83
North Ogden, UT 84414
Phone: (801) 644-0903
www.saltflats.com
Contact: Kent Singleton <kent@saltflats.com>
Meetings: 7:00 pm, first Wednesday of the month
You'll meet BYU Electric Team, WSU-EV Design Team, and other land speed racing celebrities. Always a great turnout.

Washington

Seattle Electric Vehicle Association
6021 Second Avenue NE
Seattle, WA 98115-7230
(206) 524-1351
www.seattleeva.org
Contact: Steven S. Lough <stevenslough@comcast.net>
Meetings: 7:00 pm, second Tuesday of the month

Washington D.C.

EVA of Washington DC
9140 Centerway Road
Gaitherburg, MD 20879-1882
Phone: (301) 869-4954
www.evadc.org
Contact: David Goldstein <goldie.ev1@juno.com>
Meetings: 7:00 pm, second or third Tuesday of the month

Wisconsin

Southern Wisconsin EV Proliferation
808 Fieldcrest Court
Watertown, WI 53511
Phone: (920) 261-7057
www.emissionsfreecars.com
Contact: Mike Turner <mike.turner@emissionsfreecars.co>
Meetings: Please contact for date and location

Electric Utilities and Power Associations

Any of the following organizations can provide you with information.

American Public Power Association
2301 M Street, N.W.
Washington, DC 20202
Phone: (202) 775-8300

Arizona Public Service Company
P.O. Box 53999
Phoenix, AZ 85072-3999
Phone: (602) 250-2200

California Energy Commission
1516 9th Street
Sacramento, CA 95814
Phone: (916) 654-4001

Electric Power Research Institute
412 Hillview Avenue
P.O. Box 10412
Palo Alto, CA 94303
Phone: (415) 855-2580

Director of Electric Transportation
Department of Water and Power City of Los Angeles
111 N. Hope Street, Room 1141
Los Angeles, CA 90012-2694
Phone: (213) 481-4725

Public Service Co. of Colorado
2701 W. 7th Avenue
Denver, CO 80204
Phone: (303) 571-7511

Sacramento Municipal Utility District
P.O. Box 15830
Sacramento, CA 95852-1830
Phone: (916) 732-6557

Southern California Edison
2244 Walnut Grove Avenue, P.O. Box 800
Rosemead, CA 91770
Phone: (818) 302-2255

Government

The following are agencies are involved with EVs directly or indirectly at city, state, or federal government levels.

California Air Resources Board
1012 Q Street
P.O. Box 2815
Sacramento, CA 95812
Phone: (916) 322-2990

Environmental Protection Agency
401 M Street S.W.
Washington, DC 20460
Phone: (202) 260-2090

National Highway Traffic Safety Administration
400 7th Street S.W.
Washington, DC 20590
Phone: (202) 366-1836

New York Power Authority
123 Main Street
White Plains, NY 10601

New York State Energy Research and Development Authority (NYSERDA)
17 Columbia Circle
Albany, New York 12203-6399
www.nyserda.org

Manufacturers, Converters, and Consultants

There is a sudden abundance of people and firms doing EV work. This category is an attempt to present you with the firms and individuals from whom you can expect either a completed EV or assistance with completing one.

Manufacturers

This category includes the household names plus the major independents you already met in Chapters 3 and 4. When contacting the larger companies, it is best to go through the switchboard or a public affairs person who can direct your call after finding out your specific needs.

A123Systems
The Arsenal on the Charles
321 Arsenal Street
Watertown, MA 02472
Phone: (617) 778-5700
www.a123systems.com

A123Systems Hymotion™
Phone: (877) 2GO-PHEV (877-246-7438)
www.a123systems.com/hymotion/get_charged/where_can_i_get_one
Hours: 10 am to 6 pm EST

Ampmobile® Conversions LLC
P.O. Box 5106
Lake Wylie, SC 29710
Phone: (803) 831-1082 or toll free (866) 831-1082
Email: info@ampmobiles.com

Cloud Electric Vehicles Battery Powered Systems
102 Ellison Street, Unit A
Clarksville, GA 30523
Phone: (866) 222-4035

eDrive Systems
www.edrivesystems.com

ElectroAutomotive
P.O. Box 1113-W
Felton, CA 95018-1113
Phone: (831) 429-1989
Fax: (831) 429-1907
Email: electro@cruzio.com

Energy CS
Monrovia, CA 91016
Phone: (626) 622-7376
Fax: (626) 303-7226
www.energycs.com
Contact: Mike Thompsett

EV Parts.com
108-B Business Park Loop
Sequim, WA 98382
Phone: (360) 582-1271
Phone: (888) 387-2787
Fax: (360) 582-1272
Email: sales@evparts.com

Hybrids Plus
3245 Prairie Avenue
Boulder, CO 80301
Phone: (303) 444-0569

Manzanita Micro
P.O. Box 1774
Kingston, WA 98346
Phone: (360) 297-1660
www.manzanitamicro.com
Contact: Rich Rudman

Metric Mind Engineering
9808 SE Derek Court
Happy Valley, OR 97806-7250
Phone: (503) 680-0026
Fax: (503) 774-4779
Email: ac@metricmind.com

Conversion Specialists

In this category, the line between those who provide parts and those who provide completed vehicles is blurred.

Grassroots Electric Vehicles
1918 South 34th Street
Fort Pierce FL 34947
Phone: (772) 971-0533

Vehicles and Components

Electric Transportation Applications
P.O. Box 10303
Glendale, AZ 85318
Phone: (602) 978-1373
Contact: Don Karner

Consultants

Companies and individuals who are more likely to provide advice, literature, or components—rather than completed vehicles—are listed.

Aerovironment
P.O. Box 5031
Monrovia, CA 91017-7131
Phone: (818) 359-9983
Developers of the GM Impact, Paul McCready and Aerovironment need no further introduction. In September 2007 Paul passed away after a short illness, just after retiring from Aerovironment. His insight initially sparked the concept car the GM made into the EV-1, the subject of the 2006 movie *Who Killed the Electric Car?*

Electro Automotive
P.O. Box 1113
Felton, CA 95018-1113
Phone: (831) 429-1989
Michael Brown and Shari Prange
This organization, an experienced participant in the EV field, offers books, videos, seminars, consulting, and components. Mike and Shari still supply kits for conversion builders, complete parts, and instruction manuals and are finding that with the high gasoline prices since Hurricane Katrina came ashore in 2005 that their business is brisk. They carry AC drive systems from Azure Dynamics (formerly Solectria, founded by MIT students).

Howard G. Wilson
2050 Mandeville Canyon Road
Los Angeles, CA 90049
Phone: (310) 471-7197
Former Hughes vice-president, Howard Wilson was the real "make it happen" factor behind GM's Impact and Sunraycer projects.

Bob Wing
P.O. Box 277
Inverness, CA 94937
Phone: (415) 669-7402

Suppliers

This category includes those from whom you can obtain complete conversion kits (all the parts you need to build your own EV after you have the chassis); conversion plans; and suppliers specializing in motors, controllers, batteries, chargers, and other components.

You can find more information about conversions and components at www.eaaev.org/eaalinks.html.

Amberja Projects
Unit 14, Henry Bells Yard
Dysart Road
Grantham, Lincolnshire, England
Contact: Simon Sheldon, Managing Director <simon@amberjaprojects.com>

Battery Powered Systems
204 Ellison Street, Unit A
Clarkesville, GA 30523
www.beepscom.com

EV Parts, Inc.
160 Harrison Road #7
Sequim, WA 98382
www.evparts.com/firstpage.php
They are a great component supplier.

EV Source LLC
19 W Center, Suite 201
Logan, UT 84321
Phone: (877) 215-6781
sales@evsource.com

The Green Car Company
345–106th Avenue NE
Bellevue, WA 98004
Phone: (425) 820-4549
www.thegreencarco.com

Manzanita Micro EV components
5718 Gamblewood Rd NE
Kingston, WA 98346
Phone: (360) 297-7383
Production Shop: (360) 297-1660
Metal Shop: (360) 297-3311
www.manzanitamicro.com
Contact: Rich Rudman

Provides PHEV Components and Kits

eLithion
Iris Avenue, Suite 110
Boulder, CO 80301-1956
Phone: (303) 413-1500

EV Parts, Inc.
160 Harrison Road, #7
Sequim, WA 98382
Phone: (888) 387-2787
www.evparts.com
sales@evparts.com

Metric Mind Corporation
9808 SE Derek Court
Happy Valley, Oregon 97086
Phone: (503) 680-0026
Fax: (503) 774-4779 (fax)
www.metricmind.com
Contact: Victor Tikhonov

Motors

A considerable number of companies manufacture electric motors. The short list here is only to get you started.

Advanced D.C. Motors, Inc.
219 Lamson Street
Syracuse, NY 13206
Phone: (315) 434-9303

Aveox Inc.
2265A Ward Avenue
Simi Valley, CA 93065
Phone: (805) 915-0200
www.aveox.com/Default.aspx

Azure Dynamics
9 Forbes Road
Woburn, MA USA 01801-2103
Phone: (781) 932-9009
Fax: (781) 932-9219

NetGain Technologies, LLC
900 North State Street, Suite 101
Lockport, Illinois 60441
Phone: (630) 243-9100
Fax: (630) 685-4054
www.go-ev.com
NetGain Technologies, LLC is the exclusive worldwide distributor of WarP™, ImPulse™, and TransWarP™ electric motors for use in electric vehicles and electric vehicle conversions. These powerful electric motors may also be used in the conversion of conventional internal combustion engine vehicles to hybrid gas/electric or electricassist vehicles. Their motors are manufactured in Frankfort, Illinois by Warfield Electric Motor Company.

UQM Technologies
7501 Miller Drive
P.O. Box 439
Frederick, Colorado 80530
Phone: (303) 278-2002
Fax: (303) 278-7007
www.uqm.com
UQM Technologies, Inc. is a developer and manufacturer of power-dense, high efficiency electric motors, generators, and power electronic controllers for the automotive, aerospace, medical, military, and industrial markets.

Controllers

A considerable number of companies manufacture controllers; again, this short list is only to get you started.

AC Propulsion, Inc.
462 Borrego Court, Unit B
San Dimas, CA 91773
Phone: (714) 592-5399
AC Propulsion's AC EV drive systems are simply a better idea whose time will come when economies of scale drive prices down. You'll meet them in Chapter 8. They made the tZero, two-seater sports car, and were able to prove 300 miles at freeway speeds on a single charge using commodity lithium batteries successfully.

Curtis PMC
6591 Sierra Lane
Dublin, CA 94568
Phone: (510) 828-5001
Specialists in controllers for EV applications, this company's controller is featured in Chapter 7 and used in Chapter 10's conversion. These analog DC controllers are moderately priced, provide moderate performance, and are moderately good.

eLithion
Iris Avenue, Suite 110
Boulder, CO 80301-1956
Phone: (303) 413-1500

Batteries

Alco Battery Co.
2980 Red Hill Avenue
Costa Mesa, CA 92626
Phone: (714) 540-6677
Offers a full line of lead-acid batteries suitable for EVs.

Concorde Battery Corp.
2009 W. San Bernadino Road
West Covina, CA 91760
Phone: (818) 962-4006
Offers lead-acid batteries for aircraft use.

Eagle-Picher Industries
P.O. Box 47
Joplin, MO 64802
Phone: (417) 623-8000
Offers a full line of lead-acid batteries suitable for EVs.

eLithion
Iris Avenue, Suite 110
Boulder, CO 80301-1956
Phone: (303) 413-1500
www.elithion.com

Trojan Battery Co.
12380 Clark Street
Santa Fe Springs, CA 90670
Phone: (800) 423-6569
Phone: (213) 946-8381
Phone: (714) 521-8215
Trojan has manufactured deep-cycle, lead-acid batteries suitable for EV use longer than most companies, and has considerable expertise.

U.S. Battery Manufacturing Co.
1675 Sampson Avenue
Corona, CA 91719
Phone: (800) 695-0945
Phone: (714) 371-8090
Manufactures deep-cycle, lead-acid batteries suitable for EVs. Thriving today with distributors all around.

Valence Technology
12303 Technology Boulevard, Suite 950
Austin, Texas 78727
Phone (888) VALENCE or (512) 527-2900
Fax: (512) 527-2910
www.valence.com

Yuassa-Exide
9728 Alburtis Avenue
P.O. Box 3748
Santa Fe Springs, CA 90670
Phone: (800) 423-4667
Phone: (213) 949-4266
Manufactures deep-cycle, lead-acid batteries suitable for EVs. Still making flooded lead-acid batteries, manufactured in Japan and probably China too.

Chargers

There are many battery charger manufacturers; this short list is only to get you started.

Delta-Q Technologies Corp.
Unit 3, 5250 Grimmer Street
Burnaby, BC, Canada V5H 2H2

eLithion
Iris Avenue, Suite 110
Boulder, CO 80301-1956
Phone: (303) 413-1500

Avcon Corporation
4640 Ironwood Drive
Franklin, WI 53132
Phone: (877) 423-8725
Fax: (414) 817-6161
Email: powerpak@webcom.com

Manzanita Micro
5718 Gamblewood Road NE
Kingston, WA 98346
Phone: (360) 297-7383
Contact: Rich Rudman
Designer Joe Smalley and protégé Rich Rudman make a range of fully powered factor-corrected chargers that deliver 20–50 amps DC into traction packs from 48 to 312 volts, from any line source (120 to 240 volts). These units provide remarkable power and have done well in the conversion market. Improvements (such as increasing the switching frequency) would allow a second generation to be made and packaged in a much smaller container, thereby further increasing its popularity.

News

"Political Push for Plug-in Hybrids," HybridCars.com, June 10, 2008
"Doctors Orders: Buy a Hybrid," *Environmental News Network*, June 11, 2008
"Plug-in Car Production Race Is On," *Chicago Tribune*, June 18, 2008
"Uncle Sam Rolls in a 100-mpg Solar Plug-in Hybrid," *Wired*, June 27, 2008
"Plug-in Cars Give Owners a Real Jolt of Satisfaction," *Seattle Post-Intelligencer*, September 9, 2008
"The Plug-in Revolution: A Grand Plan for America's Energy Woes," *Washington Monthly*, October 2008
Shackett, S.R. *The Complete Book of Electric Vehicles*. Domus Books, 1979.
Traister, R.J. *All About Electric & Hybrid Cars*. Tab Books, 1982.

Wakefield, E.H. *The Consumer's Electric Car*. Ann Arbor Science, 1977.
Whitener, B. *The Electric Car Book*. Love Street Books, 1981.

Manuals

Brown, M., with S. Prange. *Convert It*. Electro Automotive, 1989.
Chan, D. and K. Tenure. *Electric Vehicle Purchase Guidelines Manual*. EVAA, 1992.
Ellers, C. *Electric Vehicle Conversion Manual*. Self, 1992.
Staff, G. *Electric Car Conversion Book*. Solar Electric Engineering, 1991.
Williams, B. *Guide to Electric Auto Conversion*. Williams Enterprises, 1981.

Articles

"Battery and Electric Vehicle Update," *Automotive Engineering*, September 1992, p. 17.
Brown, S.F. "Chasing Sunraycer Across Australia," *Popular Science*, February 1988, p. 64.
Cogan, R. "Electric Cars: The Silence of the Cams," *Motor Trend*, September 1991, p. 71.
Frank, L. and D. McCosh. "Power to the People," *Popular Science*, August 1992, p. 103.
——. "Alternate Fuel Follies," *Popular Science*, July 1992, p. 54.
——. "Electric Vehicles Only," *Popular Science*, May 1991, p. 76.
Freedman, D.H. "Batteries Included," *Discover*, March 1992, p. 90.
Krause, R. "High Energy Batteries," *Popular Science*, February 1993, p. 64.
McCready, P. "Design, Efficiency and the Peacock," *Automotive Engineering*, October 1992, p. 19.
Meyers, P.S. "Reducing Transportation Fuel Consumption," *Automotive Engineering*, September 1992, p. 89.
Pratt, G.A. "EVs: On the Road Again," *Technology Review*, August 1992, p. 50.
"Propulsion Technology: An Overview," *Automotive Engineering*, July 1992, p. 29.
White, D.C., et al. "The New Team: Electricity Sources Without Carbon Dioxide," *Technology Review*, January 1992, p. 42.

Publishers

Here are a few companies that specialize in publications of interest to EV converters.

Battery Council International
401 N. Michigan Avenue
Chicago, IL 60611
Phone: (312) 644-6610
Publishes battery-related books and articles.

Institute for Electrical and Electronic Engineers (IEEE)
IEEE Technical Center
Piscataway, NJ 08855
Publishes numerous articles, papers, and proceedings. Expensive, but one of the best sources for recent published technical information on EVs.

Lead Industries Association
292 Madison Avenue
New York, NY 10017
Publishes information on lead recycling.

Society of Automotive Engineers (SAE) International
400 Commonwealth Drive
Warrendale, PA 15096-0001
Phone: (412) 772-7129
Publishes numerous articles, papers, and proceedings. Also expensive, but the other best source for recent published technical information on EVs.

Newsletters

Here are a few companies that specialize in newsletter-type publications of interest to EV converters.

Electric Grand Prix Corp.
6 Gateway Circle
Rochester, NY 14624
Phone: (716) 889-1229

Electric Vehicle Consultants
327 Central Park West
New York, NY 10025
Phone: (212) 222-0160

Solar Mind
759 S. State Street, #81
Ukiah, CA 95482
Phone: (707) 468-0878
Electric vehicle directories.

Online Industry Publications

These online publications report on industry activity, manufacturer offerings, and local electric drive–related activities.

Advanced Battery Technology
www.7ms.com/abt/index.html

Advanced Fuel Cell Technology
www.7ms.com/fct/index.html

Earthtoys
www.earthtoys.com
A resource for alternative energy and hybrid transportation information and features. In addition to the bimonthly emagazine, there is also an up-to-date news page, link library, company directory, event calendar, product section, and more.

e-Drive Magazine
www.e-driveonline.com
Features new products, services, and technologies in motors, drives, controls, power, electronics, actuators, sensors, ICs, capacitors, converters, transformers, instruments, temperature control, packaging, and all related subsystems and components for electrodynamic and electromotive systems.

Electrifying Times
www.electrifyingtimes.com
Provides interesting information on electric vehicles and the industry.

EV World
evworld.com
Houses an online "library" of EV-related reports, articles, and news releases available to the general public. EV owners also can register and share their experiences with others. Visitors can sign up for a weekly EV newsletter. EV World has information about conversions, conversion suppliers, and a list of popular EV conversion vehicles (www.evworld.com/archives/hobbyists.html).

Fleets & Fuels
www.fleetsandfuels.com
A biweekly newsletter (distributed online) providing business intelligence on alternative fuel and advanced vehicles technologies encompassing electric drive, natural gas, hydraulic hybrids, propane and alcohol fuels, and biofuels. The newsletter is dedicated to making the AFVs business case to fleets.

Greencar Congress
www.greencarcongress.com

Hybrid & Electric Vehicle Progress
www.hevprogress.com
Formerly Electric Vehicle Progress. Follows new EV products, including prototype vehicles; provides status reports on R&D programs; publishes field test data from

demonstration programs conducted around the world; details infrastructure development, charging sites, and new technologies; and includes fleet reports, battery development, and a host of other EV-related news. Published twice a month.

Industrial Utility Vehicle & Mobile Equipment Magazine
www.specialtyvehiclesonline.com
Dedicated to engineering, technical, and management professionals as well as dealers and fleet managers involved in the design, manufacture, service, sales, and management of lift trucks, material handling equipment, facility service vehicles and mobile equipment, golf cars, site vehicles, carts, personal mobility vehicles, and other types of special purpose vehicles.

Federal Government Sites

IRS Forms—EV Tax Credits
Qualified EV Tax Credit Forms must accompany any tax returns that are claiming the ownership or purchase of a qualified EV.

Advanced Vechicle Testing Program
www.avt.inel.gov
Office of Transportation Technologies, U.S. Department of Energy. This website is run by the Idaho National Engineering and Environmental Laboratory (INEEL). It offers EV fact sheets, reports, performance summaries, historical data, and a kids' page. Visitors can also request information online.

Alternative Fuels Data Center
www.eere.energy.gov/cleancities
A comprehensive source of information on alternative fuels. Sections include an interactive map of AFV refueling stations in the U.S.; listings and descriptions of different alternative fuels and AFV vehicles; online periodicals; and resources and documents on AFV programs. The site is part of the National Renewable Energy Laboratory's (NREL) website.

Energy Information Administration
www.eia.doe.gov

NREL Home Page
www.nrel.gov
The National Renewable Energy Laboratory (NREL) has created a website detailing research efforts in renewable energies and alternative transportation technologies. Some key areas include hybrid vehicle development, renewable energy research, and battery technology research.

Office of Transportation Technologies EPAct & Fleet Regulations
www.eere.energy.gov/vehiclesandfuels/epact/
Many public and private fleets are subject to AFV acquisition requirements under the Energy Policy Act (EPAct) regulations. These requirements differ for different types of fleets. Visit this site to obtain information on fleet requirements and the manners in which you can comply with the EPAct regulations.

The US Department of Defense Fuel Cell Program
www.dodfuelcell.com

Thomas
www.congress.gov
Acting under the directive of the leadership of the 104th Congress to make federal legislative information freely available to the Internet public, a Library of Congress team brought the THOMAS World Wide Web system online in January 1995. The THOMAS system allows the general public to search for legislation and information regarding the current and past business of the U.S. Congress.

US Department of Transportation Advanced Vehicle Technologies Program
http://scitech.dot.gov/partners/nextsur/avp/avp.html
The homepage includes links to regional members of the Advanced Vehicle Program (AVP):

- **Mid-Atlantic Regional Consortium for Advanced Vehicles (MARCAV)**
  www.marcav.ctc.com
  A Pennsylvania-based organization that was established to organize industrial efforts to develop enhanced electric drives for military, industrial, and commercial vehicles. Visitors can review a list of MARCAV projects and research specific projects.
- **Hawaii Electric Vehicle Demonstration Project**
  www.htdc.org/hevdp
  A consortium dedicated to furthering electric vehicle development and sales in Hawaii. The site provides visitors with background on the program and lists accomplishments.
- **Center for Transportation and the Environment (CTE)**
  www.cte.tv
  CTE is a Georgia-based coalition of over 65 businesses, universities, and government agencies dedicated to researching and developing advanced transportation technologies. The site includes industry news, studies and projects, a database of products, and a section on EV education.
- **CALSTART/WestStart**
  www.calstart.org
  A California-based nonprofit organization dedicated to "transforming transportation for a better world." Visitors can read daily and archived

industry news updates and publications, search EV-related databases, and interact with other EV owners in an online forum.

- **Northeast Alternative Vehicle Consortium (NAVC)**
  www.navc.org
  A Boston-based association of private and public sector organizations that works to promote advanced vehicle technologies in the Northeast. Visitors can read about NAVC projects and link to related Internet sites.

State- and Community-Related Electric Vehicle Sites

California Air Resources Board (CARB)

www.arb.ca.gov
This site provides access to information on a variety of topics about California air quality and emissions. The site has general information on all types of alternative-fueled vehicle programs and demonstrations. The CARB's mission is to promote and protect public health, welfare, and ecological resources through the effective and efficient reduction of air pollutants while recognizing and considering the effects on the economy of the state.

The California Air Resources Board's guide to zero and near zero emission vehicles is available at Driveclean.ca.gov.

California Energy Commission

www.energy.ca.gov
This site gives viewers access to information on a variety of topics about California's energy system. The site dedicates a page to electric vehicles, where it has general information on electric transportation, lists sellers of EDs in California, outlines state and federal government incentives for AFVs, and includes a database of contacts in the electric transportation industry.

Mobile Source Air Pollution Reduction Review Committee (MSRC)

www.msrc-cleanair.org
The MSRC was formed in 1990 by the California legislature. The MSRC website offers information on a variety of topics regarding California air quality and programs underway to improve it, including a number of EV-related programs and incentives.

San Bernardino Associated Governments (SANBAG)

www.sanbag.ca.gov
SANBAG is the Council of Governments and Transportation Commission for San Bernardino County. The site has various information about current transportation projects underway in the San Bernardino area, as well as information for commuters Further, the site contains funding alerts for individuals and companies looking to obtain project funding and/or assistance.

General Electric Drive Information Sites

Many web sites disseminate information on EDs or report industry news and developments. A few of these, which "house" specific EV-related information, are provided here.

Advanced Transportation Technology Institute

www.atti-info.org
The Advanced Transportation Technology Institute (ATTI), a nonrofit organization, promotes the design, production, and use of battery-powered electric and hybridelectric vehicles. The organization supports individuals and organizations interested in learning more about electric and hybrid-electric vehicles, particularly electric buses.

Alternative Fuel Vehicle Institute

www.afvi.org/electric.html
AFVI was formed by Leo and Annalloyd Thomason, who each have more than 20 years' experience in the alternative fuels industry. In 1989, following more than five years' natural gas vehicle market development work for Southwest Gas Corporation and Lone Star Gas Company, the Thomasons founded Thomason & Associates. The company quickly became a nationally known consulting firm that specialized in the market development and use of alternative transportation fuels, particularly natural gas. In this capacity, they incorporated the California Natural Gas Vehicle Coalition and worked extensively with the California Legislature, the California Air Resources Board, the South Coast Air Quality Management District, and other government agencies to establish policies and programs favorable towards alternative fuels. Thomason & Associates also conducted market research and analyses, developed dozens of alternative fuel vehicle (AFV) business plans, and assisted clients in creating markets for their AFV products and services.

Association for Electric and Hybrid Vehicles

www.asne.nl/
ASNE is the Dutch division of the Association Européenne des Véhicules Electriques Routiers (AVERE), an association founded under the auspices of the European Community. The goal of ASNE is to encourage the easy use of totally or partly (hybrid) electric vehicles and vehicles with other alternative propulsion systems in road traffic.

Blogs

New Energy News

www.newenergynews.blogspot.com
Renewable energy news.

Plugs and Cars

www.plugsandcars.blogspot.com
About plug-in vehicles.

RechargeIT

www.rechargeit.blogspot.com
"Recharge a Car, Recharge the Grid, Recharge the Planet."

ZEVInfo

www.zevinfo.com
Designed by the California ZEV Education and Outreach Group, which was established under the California Air Resources Board's (CARB) ZEV Program. The basis of the website is to serve as a "one-stop-shop" for information on electric drive products in California. Moreover, the website's goal is to inform the public of the benefits and availability of advanced electric drive technologies, from early deployment and on into the future.

Other Related Websites

Canadian Environment Industry Association

www.ceia-acie.ca
The Canadian Environment Industry Association (CEIA) is the national voice of the Canadian environment industry. CEIA is a business association that, along with its provincial affiliates, represents the interests of 1,500 companies providing environmental products, technologies, and services.

Fair-PR

www.fair-pr.com/background/about.php
The largest international commercial exhibition on hydrogen and fuel cells at the Hannover Fair in Germany, featuring over 100 companies and research institutions from 30 countries.

National Station Car Association

www.stncar.com
Although closed at the end of 2004, the National Station Car Association worked for 10 years to guide the development and testing of the concept of using battery-powered cars for access to and egress from mass transit stations, and to make mass transit a convenient door-to-door service. The NSCA released a report (National Station Car Association History) that gives an overview of the program's history.

Notes

Chapter 1

1. www.en.wikipedia.org/wiki/Plug-in_hybrid#cite_note-hevctr-0.
2. www.geosci.uchicago.edu/~archer/reprints/archer.2005.fate_co2.pdf.
3. Bob Brant and Seth Leitman, *Build Your Own Electric Vehicle*, 2nd ed. (New York: McGraw-Hill, 2008).
4. Ibid.
5. Hybridcars.com.
6. A. Frank et al., "What Are Plug-In Hybrids?" Team Fate (University of California, Davis); retrieved August 7, 2007; earlier version.
7. E. Knipping and M. Duvall, M. (June 2007) *Environmental Assessment of Plug-In Hybrid Electric Vehicles*, Volume 2, *United States Air Quality Analysis Based on AEO-2006 Assumptions for 2030*, Electric Power Research Institute and Natural Resources Defense Council; accessed July 21, 2007.
8. Wikipedia.
9. Calcars.org.
10. Michael d'Estries, Groovy Green.
11. Brant and Leitman.

Chapter 2

1. www.en.wikipedia.org/wiki/Energy_policy_of_the_United_States#_note-PetFact; U.S. Department of Energy, "Basic Petroleum Statistics: Energy Information."
2. www.en.wikipedia.org/wiki/Energy_policy_of_the_United_States#_note-10#_note-10;www.finance.senate.gov/hearings/testimony/2007test/050107testwm.pdf, 2006; retrieved July 4, 2007.
3. Wang, DeLuchi, and Sperling.

4. Kenneth S. Kurani, Reid R. Heffner, and Thomas S. Turrentine, "Driving Plug-In Hybrid PHEVs: Reports from U.S. Drivers of HEVs converted to PHEVs, circa 2006-07," Plug-in Hybrid Electric Vehicle Research Center, Institute of Transportation Studies, University of California, Davis, funded by University of California Energy Institute and PIER Plug-in Hybrid Electric Vehicle Research Center, October 16, 2007.
5. Ibid.
6. Clean Air Act of 1990 amendments.
7. Courtesy of Wikopedia

Chapter 3

1. "Hybrid Car Ready in 1969," *Popular Science*, July 1969, pp. 86–87.
2. http://www.popsci.com/cars/article/2002-02/persistent-contender-popsci-covers-hybrids.
3. Bob Brant and Seth Leitman, *Build Your Own Electric Vehicle*, 2nd ed. (New York: McGraw-Hill, 2008).
4. www.hybridcars.com/history/history-of-hybrid-vehicles.html.
5. Ibid.
6. See "How Carmakers Are Responding to the Plug-In Hybrid Opportunity," CalCars; www.calcars.org/carmakers.html.
7. SAE 971629, "Early SAE PHEV Paper."
8. CalCars.org.
9. www.inoculatedmind.com/2006/10/episode-33/.
10. www.eaa-phev.org/wiki/PriusPlus.
11. http://www.calcars.org/photos.html.
12. PRIUS+.
13. http://en.wikipedia.org/wiki/Plug_In_America.
14. www.calcars.org/history.html.
15. Green Car Congress, "Advanced Hybrid Vehicle Development Consortium Targets Plug-Ins."
16. Wikipedia.
17. *Jay Friedland, Plug In America;* www.californiaprogressreport.com/2009/02/stimulus_bill_b.html.
18. blog.cleveland.com/business/2008/10/bailout_bill_includes_tax_break.html.
19. www.motorauthority.com/senate-passes-bill-that-approves-up-to-7500-in-tax-credits-for-plug-ins.html.
20. CalCars.org.
21. www.nyserda.org.
22. www.a123systems.com/company.
23. www.greencarcongress.com/2007/05/a123systems_to_.html#more.
24. Ibid.

25. CalCars.org.
26. www.electrifyingtimes.com/GM_Volt/electric_cars_and_the_volt.html.
27. The Daily Green; www.thedailygreen.com/living-green/blogs/cars-transportation/green-cars-detroit-auto-show-46011608.
28. Ibid.
29. www.greencarcongress.com/2007/05/a123systems_to_.html#more.
30. www.calcars.org/vehicles.html.
31. Sherry Boschert, *Plug-in Hybrids: The Cars That Will Recharge America*, New Society Publishers, 2006, p. 5.

Chapter 4

1. AFS Trinity.
2. AFS Trinity; www.afstrinity.com/press-images.htm.
3. www.hipadrive.com/Documents/pressrelease.pdf.
4. Sherry Boschert, *Plug-in Hybrids: The Cars That Will Recharge America*, p. 76.
5. Sara Knight, "Green City: Plugging Into What's Next," *San Francisco Guardian* Online, October 9, 2007.
6. www.eaa-phev.org/wiki/PriusPlus.

Chapter 5

1. www.afdc.energy.gov/afdc/fueleconomy/animations/hybrids/swfs/hybridframe.html.
2. www.afdc.energy.gov/afdc/fueleconomy/animations/hybrids/hybrid/hybridoverview.html.
3. www.afdc.energy.gov/afdc/vehicles/hybrid_electric_series.html.
4. www.afdc.energy.gov/afdc/vehicles/hybrid_electric_parallel.html.
5. Wikipedia.
6. www.afdc.energy.gov.
7. Bob Brant and Seth Leitman, *Build Your Own Electric Vehicle*, 2nd ed. (New York: McGraw-Hill, 2008).
8. U.S. Department of Energy.

Chapter 6

1. www.en.wikipedia.org/wiki/Hybrid_electric_vehicle#cite_note-21.
2. www.en.wikipedia.org/wiki/Hybrid_electric_vehicle#cite_note-22.
3. www.eaa-phev.org/wiki/IMA.
4. Ibid.
5. www.en.wikipedia.org/wiki/History_of_plug-in_hybrids#cite_note-6#cite_note-6.
6. www.eaa-phev.org/wiki/Insight_PHEV.

Chapter 7

1. Elithion; www.elithion.com.
2. www.liionbms.com/php/about_bms.php.
3. Ibid.
4. www.liionbms.com/php/controllers.php#HV%20front%20ends.
5. Elithion; www.eLithion.com.
6. EAA-PHEV.
7. Ibid.
8. Elithion; www.eLithion.com.
9. EAA-PHEV

Chapter 8

1. Bob Brant and Seth Leitman, *Build Your Own Electric Vehicle*, 2nd ed. New York: McGraw-Hill, 2008.
2. www.eaa-phev.org/wiki/PriusPlus.
3. Andrew Burke and Eric Van Gelder, *Plug-in Hybrid-Electric Vehicle Powertrain Design and Control Strategy Options and Simulation Results with Lithium-ion Batteries*, Institute of Transportation Studies, University of California–Davis, One Shields Ave., EET-2008, European Ele-Drive Conference, International Advanced Mobility Forum, Geneva, Switzerland, March 11–13, 2008.
4. www.energy.gov/news/5523.htm.
5. Brant and Leitman.
6. www.eaa-phev.org/wiki/PriusPlus.
7. CalCars.
8. www.eaa-phev.org/wiki/PriusPlus.

Chapter 9

1. www.eaa-phev.org/wiki/Prius_PHEV_Battery_Options.
2. www.manzanitamicro.com/hybrids.htm.
3. www.liionbms.com/php/chargers.php.
4. Bob Brant and Seth Leitman, *Build Your Own Electric Vehicle*, 2nd ed. New York: McGraw-Hill, 2008.
5. www.liionbms.com/php/about_charge_control.php.
6. Elithion.
7. www.manzanitamicro.com/hybrids.htm.
8. EAA-PHEV,
9. Brant and Leitman, Chapter 9.
10. EAA-PHEV.
11. www.eaa-phev.org/wiki/Escape_PHEV_TechInfo.
12. www.eaa-phev.org/wiki/Escape_PHEV_TechInfo#HV_connector:_C4227C.
13. www.eaa-phev.org/wiki/Escape_PHEV_TechInfo.
14. Ibid.

Chapter 10

1. Bob Brant and Seth Leitman, *Build Your Own Electric Vehicle*, 2nd ed. New York: McGraw-Hill, 2008.
2. Ron Gremban, 2004, update in 1/18/2006.
3. EAA-PHEV.
4. Ibid.
5. Currently used in Ron Gremban's Toyota Prius PHEV. www.eaa-phev.org/wiki/PriusPlus_History#Appendix_A:_Battery_Specifications.
6. EAA-PHEV.
7. Ibid.
8. Ibid.
9. www.eaa-phev.org/wiki/CAN-View
10. www.eaa-phev.org/wiki/PiPrius_conversion_process#_note-0.
11. EAA-PHEV.

Chapter 11

1. Wikipedia.
2. www.liionbms.com/php/phev_faq.php#toc1.
3. Irv Miller, Vice President—Corporate Communications, Toyota Motor Sales Group, Toyota Open Road blog.
4. www.liionbms.com/php/phev_faq.php#toc56.

Index

A

B

D

G

H

L

M

N

O

T

100+ MPG
Hybrids!

www.calcars.org

better, cleaner.

I'm a Charter Sponsor of the California Cars Initiative

AN ONLINE FORUM FROM THE GREEN LIVING GUY

www.greenlivingguy.com

Deeply rooted within the music and environmental communities, Reverb educates and engages musicians and their fans to take action toward a more sustainable future.

Reverb works with artists to minimize the carbon footprint associated with touring by implementing both front stage and backstage greening elements. Some of these backstage components include coordinating biodiesel fueling for tour buses and trucks, setting up extensive recycling programs, providing biodegradable catering products and non-toxic cleaning supplies, and setting the band and crew up with customized water bottles.

Front of stage components are aimed at reaching fans and encouraging them to take action. Informative greening websites let fans know about the green steps their favorite artists are taking and provide resources like online carpooling networks, volunteer opportunities, and more. Reverb's Eco-Village, set up before each concert, allows fans to offset their carbon footprint, sample eco-friendly products, and learn more about actions they can take in their own lives.

With Reverb's help, artists are able to use their voice and set an example to encourage fans to be more mindful of the environment.

- Virtual Eco-Village/Mini-Site
- Facebook
- Myspace
- E-blast content
- Online Outreach
- Carbon Neutral Concerts and Venues
- Biodiesel for Vehicles and Generators
- Waste Reduction
- Biodegradable Catering Products
- Recycling
- Green Bus Supplies and Cleaners
- Energy Efficiency
- Green Contract Rider
- Eco-Friendly Merchandise
- Green Sponsorship
- On Site and On-Line Fan Outreach

For more information: info@reverbrock.og | http://www.reverbrock.org

John Mayer's brand, AKOG (Another Kind of Green) was created from the belief that small steps toward environmental sustainability can effect widespread change when multiplied by a great number of participants.

In fact, through participation in the AKOG program, fans have already offset over 2,200 *tons* of CO_2 pollution, equal to not driving over 4.4 MILLION MILES OF DRIVING!

Join John Mayer in the fight against global warming and take your first step today.

- Carbon offsets to account for CO_2 emissions from venue energy use, trucks and busses, flights and hotels.
- Inviting local and national non-profit groups to be a part of the Reverb Eco-Village to educate and engage fans
- Sustainable supplies such as biodegradable and reusable catering products and local and organic food

In conjunction with Reverb, we will be helping offset each show with wind power, putting together a "village" in the concourse that consists of environmentally and socially minded non-profits and green sponsor types, and most importantly providing cool offset stickers so you can neutralize the pollution from your drive to and from the show

For more information: info@reverbrock.og | http://www.reverbrock.org